T0200315

Technical Education for Sustainability

Environmental Education, Communication and Sustainability

edited by Walter Leal Filho

Vol. 30

PETER LANG

Frankfurt am Main · Berlin · Bern · Bruxelles · New York · Oxford · Wien

Mahshid Sotoudeh

Technical Education for Sustainability

An Analysis of Needs in the 21st Century

PETER LANG
Internationaler Verlag der Wissenschaften

Bibliographic Information published by the Deutsche Nationalbibliothek
The Deutsche Nationalbibliothek lists this publication in the Deutsche
Nationalbibliografie; detailed bibliographic data is available in the internet at
<http://www.d-nb.de>.

Gedruckt mit Unterstützung des Bundesministeriums für
Wissenschaft und Forschung in Wien.

Layout: www.kumpernatz-bromann.de

ISSN 1434-3819
ISBN 978-3-631-58594-8

© Peter Lang GmbH
Internationaler Verlag der Wissenschaften
Frankfurt am Main 2009
All rights reserved.

Printed in Germany 1 2 3 4 5 7

www.peterlang.de

Preface

This book is about understanding the technical education challenges in the context of sustainable development. The study considers three different perspectives: the engineering profession view point, the perspective of technology development in a socio-technical system and the technical education view point. The study provides a reflection on the engineering profession and the services it provides in sustainable development. The author aims to use the book for introducing the engineering profession to challenges from different perspectives, for engineering students in seminars for sustainability and as a guide for students' discussions regarding their future career in a sustainable development.

Moreover, the book includes a contribution to the debates on reforms of technical universities in sustainable development and presents arguments for a participative concept for reform strategies. Finally a number of future research items are presented.

Acknowledgement

The author would like to thank Professor Michael Narodoslawsky for regular dialogues with him to shape the frame of the research, his comprehensive reviews and his valuable suggestions.

Institute of Technology Assessment supported the research conducted for this work and was the place for dialogues with, Dr. Helge Torgersen, Doz. Michael Nentwich and Professor Gunther Tichy to whom the author's debt is beyond expression.

The author would like to express her deepest thanks to all respondents to the survey and to Dr. Christina Raab who studied the very first version of the manuscript and made constructive suggestions.

The author is indebted to her parents for their valuable and unlimited support and thanks Tina for being so patient.

The author's debt is beyond redress to Dr. Vahid Sotoudeh for reviewing the final draft and his assistance and support for conveying the survey.

Content

Figures

Tables

Boxes

1. Introduction

The main question addressed in this book is "what roles should technical education and the engineering profession have in the future from a sustainable development perspective?". This book is a contribution to the discussions within the scope
of United Nations Decade of Education in Sustainable Development 2005-2014.

As an engineer, who analyses the impacts of technologies[1], the author has been
able to categorise different domains within the world of engineering. The domain
of technology assessment tells us that we do not know much about many potential
impacts of our technical design that might have serious consequences for human
beings. The domain of engineering profession and practice shows how engineers
enthusiastically work in order to design and improve the infrastructures for transportation, communication or manufacturing, and design and develop innovative
technical solutions to fulfil the needs of users and consumers. And last but not
least the domain of technical education is the root for engineering knowledge and
skills. The common features among these three domains are the role of technology in society and the contrasts among different social values against technology
development. In this book the author analyses the need for technical education
and technical universities to prepare engineers for the 21st century with an understanding of the potential impacts of technical design. These needs are discussed in
relation to social values that influence the strategies for technical education.

Universities and academies have been elements of civilization for generation
and conservation of knowledge. The ratios among the generation and conservation of knowledge and the quality, content and dynamics of these processes have
been continuously changing throughout different time periods and in different
cultures. Experimentation and practical knowledge were introduced at a later
stage to universities. Those who worked on the design and structure of buildings,
water canals, and mechanical instruments often learned and generated their
knowledge and tested them outside the universities. The modern forms of technical schools, Polytechnics, institutes and universities of technology (referred to
as technical universities collectively in this book) have the autonomy to offer
bachelors, masters or doctoral degrees and were as such first established during
the industrial revolution. Technical universities founded in the late eighteenth/
early nineteenth century in Europe were originally designed for military education and industrial-technical engineering fields such as mining and construction
engineering (see also Jischa 2004). It was the alumni of these same universities
who established and shaped the technical universities in the nineteenth century
in the Americas and later in Australia. Their ideas also served as a guide for the

---

[1] Glossary.

reforms of technical schools in the twentieth century at the older Asian universities.

Technical universities have received economic support not only for the education of engineers but also for the development of technical innovations, which have been important since World War I for the expansion of the states military and industrial power. In the last decades of the twentieth century technical universities were faced with new debates on critical issues such as:

- rapid economic changes in relation to new technologies such as information technology,
- global environmental impacts of industrial activities,
- changing working conditions and organizational structures for engineering technicians and engineers,
- new forms of learning such as e-learning.

During that time period technical universities started receiving increasing signals from external private and public communities regarding the environmental concerns of the general public. These signals originated from the industries needs to avoid accidents leading to environmental catastrophes and from massive social discontent and health problems in industrialized countries. Additionally complaints were submitted by countries affected by trans-boundary (long distance) air and/or water pollution caused by the industrial activities of other countries. Environmental and social concerns were, however, only part of the social movements at that time. Globalisation, urbanisation and individualisation on one hand and liberalisation of the market and the need for more mobility on the other hand led to the need for organisational reforms of different social systems and organisations. The knowledge-based society also generated new requirements for technical education.

After an initial reaction by technical universities to mostly environmental and economic concerns in the last decade of the twentieth century further concerns surfaced revealing that the structures of curricula, management and research at technical universities did not match the new needs. Various strategies have been created at local, national, regional and international level to plan and apply reforms:

- individual actors decided independently on desired local changes and started initiatives at individual universities,
- engineering associations, education policy makers and research funds set out declarations, programs and strategies based on their understanding of the changes needed.

Since the identified needs differ, the reform of technical education and research at technical universities has been channelled in very different directions. New strategies focus inter alia on

- adaptation of universities to new economic and business conditions through reorganisation, change of management strategies or new education programs with more practice- and business-oriented curricula[2],
- changes to the curricula so as to consider environmental issues with pioneers such as Lunenburg university in Germany[3],
- partnerships between universities and local communities (see Dyer/Holm 2006),
- consideration of sustainable development concepts in the planning and reforms at different levels and scopes for universities (see University of Florida 2005).

The impacts of these strategies on technology development and the engineering profession have not yet been investigated. Although engineers and engineering technicians (craftsmen) constitute a large occupational group in industrial countries, there are only a few studies available on the potential impacts of the different organisational changes on their education. There is also little information available on the impacts of new working conditions on engineering work and the need for new engineering skills and knowledge as well as on communication mechanisms among the relevant actors for technical education, the engineering profession and technology development at local and global levels.[4]

Available discussions on the role of engineering in society are often polarized. There are different understandings of the responsibility of the engineering profession and individual engineers for the consequences of technical solutions. In 1971 the mechanical engineer Edwin Layton called engineers both scientists and businessmen with sometimes controversial interests. In his book "The Revolt of the Engineers" he analysed situations where the engineers' acting responsibly is the only way to avoid catastrophes. On the other hand there are studies that present engineers as bored and lacking a "people orientation" with a narrow range

---

2   Establishment of Science, technology and research parks are examples of these strategies. Stanford University was one of the pioneers in these strategies: "Stanford University created the first high-tech research park in 1951 in the U.S.A. in response to the demand for industrial land near university resources and an emerging electronics industry tied closely to the School of Engineering. The first lessee of the Stanford Research Park was Varian Associates. Instrumental to the creation and growth of Silicon Valley, the park is now home to more than 140 companies in electronics, software, biotechnology and other high-tech fields." (Stanford university 2007).

3   See project examples in http://www.leuphana.de/?id=5092/pdf/Jubi-Projekte-Publika-tionen.pdf, accessed 13 May 2008.

4   (ANSB 2006) presents quantitative data on the U.S. and international science and engineering enterprise.

of interests.[5] A number of references are used for this study to identify different perspectives on roles of engineering work and success factors for engineers.[6]

Without doubt the history of engineering presents a variety of examples of the character of an engineer with different affinities to socially relevant issues. It seems that two approaches are necessary to follow to understand engineering and engineering (technical) education needs. These are:

1. asking engineers themselves to use their own vocabulary and ideas, and a study of their work mechanisms and

2. investigating conditions by social scientists who have an interest in work sociology at a micro level (see Domal/Trevelyan 2006).[7]

---

5   (Beder 1989).
6   See (Domal/Trevelyan 2006): e.g. the importance of coordination and communication activities for engineers, the differences between engineering in different cultures; (Nichol 2006): e.g. the need for broad base, systematically educated engineers, a broad understanding of sustainability principles; (O'Connell 2006): e.g. the responsibility of engineering profession: a "true" profession is a way of life in which expert knowledge is used not primarily for personal gain but for the benefit of those who need that knowledge; (Sotoudeh 2006): e.g. better understanding of the needs of the society, communication skills for participative methods; (ANSB 2005): e.g. greater career opportunities for engineers with a broader set of skills, life-long learning skills, system thinking, an ability to work in a multicultural environment, an ability to understand the business context of engineering, interdisciplinary skills, communication skills, leadership skills; (Clough 2005): need for engineers who are well grounded in the humanities, social sciences, and economics as well as science and mathematics, need for strong analytical skills, practical ingenuity, high ethical standards, ability to put problems in their socio-technical and operational context; (Dodds/Venables 2005): Look beyond their own locality and the immediate future, seek engagement from all stakeholders, adopt a holistic,'cradle-to-grave' approach; (RAE 2003): e.g. basic mathematics and physical science skills, modernisation of degree courses through the incorporation of subjects such as business and communication, promoting participation of women in engineering, need for entrepreneurial engineers; (Roman 2003): e.g. hybrid character of engineering profession to combine different kinds of science and skills (cited from M. P. O'Brien); (Long/Failing 2002): e.g. responsibility of engineers to familiarise themselves with sustainability issues, fully investigating the impact of potential actions, Weighing the impacts of alternative solutions, fostering consultation and partnership; (El-Sayed 2001): cooperative education which alternates between academic work and industrial work; (SEFI 2001): e.g. communication skills, team work, need for broad minded engineers, social intelligence, reflection, creativity; (Emblemsvag/Bras 2000): e.g. system thinking versus process thinking for design of products; (Marczyk 2000): e.g. dealing with uncertainty in engineering, Computer-Aided-Engineering; (Canel 2000): barriers to women in engineering; (Rae/Volti 1999): e.g. Role of engineers in organisations; (Beder 1989): e.g. problem-solving capability, communication skills, engineering as social activity.
7   Organizational sociology at macro level neglects the fine details in the engineering work.

Engineering journals are a good source for investigating engineers' ideas and opinions. These journals show generally that in the last decades of the twentieth century engineers became more interested in finding and implementing solutions to environmental problems. However, the intensity of engagement and the understanding of the problem differ within different engineering communities.

At the beginning of the new millennium a small group of about three hundred engineers and scientists from different disciplines started an intensive discussion on the role of engineers and technical universities in working towards sustainable development in the 21st century at an international conference in Delft. This international conference on Engineering Education in Sustainable Development (EESD) has been organized every two years since 2002. In October 2004 the conference issued the Declaration of Barcelona as an input to the discussions regarding the role of technical universities and engineers in sustainable development. This declaration shows the broad spectrum of new needs from the perspective of an engineering community (which includes technicians, too). Furthermore it suggests activities for the modification of curricula and the reorganization of campuses among others for "green universities" or "sustainable universities". One of the objectives of EESD conferences is to provide communication among all engineering disciplines and identify synergy effects among different universities, disciplines and cultures for technical education.

Having taken into account the diversity of backgrounds and roles of engineers and motivations for changes of engineering work and education and having followed the debates on sustainable universities over the past years, it can be concluded that the "future of technical education" is not only a management task for individual universities or a discussion issue for the media but also it is a complex and socially relevant issue that needs to be addressed by participation of all relevant stakeholders.

The analysis presented in this book aims at identifying interrelated factors that influence technical education through understanding of relations (system understanding) and identification of needs and objectives for technical education (target knowledge). The study comes to its conclusion in this book with some preliminary recommendations for a discussion on the future of technical universities based on the new role of engineering profession and technology in society. Questions as topics for further research are included as one of the recommendations.

In this book the discussion starts with consideration of a society where people have the opportunity to know about the risks and rewards of technical innovation and where democracy guarantees basic human rights and peaceful development not only for the present but also for the future generations. These considerations constitute the basic elements of the concept of sustainable development, namely basic human rights in a democracy and the precautionary principle for protecting the chances of future generations. Sustainable development considers technology development within the framework of democracy and peaceful activities. The term sustainable development is deliberately avoided in the first chapters to prevent misunderstandings due to the vague definition of this concept. The term is only used, when it is placed in a clear context such as the socio-political dimension of sustainable development[8] or discussions on the Declaration of Barcelona for technical education in a sustainable development (in section 4.2).

This book consists of three parts as follows:

Part I contains evidence for needs of technical education based on a number of lessons learned from history and a survey of engineers' opinions regarding the factors that influence the efficacy and success of future engineers. This survey was done by the author for this book in 2006 in order to gather ideas on factors defining "success" in different contexts for the engineering profession in the language of engineers themselves.

Part II presents evidence for defining the scope of the changes needed in technical education based on an understanding of socio-technical systems and their interrelations with engineering knowledge and praxis. In this Part the results of a literature survey have been presented in order to develop a concept for a socio-technical system and to identify the sustainable technology development challenges for the engineering profession.

One important result from the analysis in Parts I and II is the understanding of challenges, needs and success factors for peaceful engineering practice in a democratic society with commitment to the precautionary principle. Since this analysis is focused on the identification of needs for technology development and the engineering profession, it is an important element for a requirement analysis for the output of technical universities. This understanding of the engineering challenges is used in Part III in order to identify crucial questions for strategies of technical universities preparing the engineers of the future who will be expected to deal with such challenges. The normative framework of sustainable development is applied in Part III to discuss the path toward changes of technical education.

This definition of sustainable development includes, as mentioned earlier, basic human rights in a democracy and the precautionary principle for protecting the chances of future generations. Suggestions for the paths towards changes are made after presenting a discussion of the socio-political dimension of sustainable development. The resulting preliminary recommendations serve as the starting point for future research on the need for changes in technical education.

---

8   See also Albert et al. 2001.

2. Background to the methods applied in the study

This book focuses on the engineering profession as a service to society for civil and peaceful applications. This interpretation also implies minimizing risks to society through applying the precautionary principle when technical solutions have a high risk potential and there are considerable uncertainties on the impacts of the solutions (see Sanderson et al. 2002). A special normative background in Part III of this book addresses the short- and long-term needs of human beings according to the principles of sustainable development.

Challenges faced by technical education in the 21st century, to provide skills and expertise for the engineering profession, are identified in this book from a broad social perspective. Technical education is analysed in the context of engineering work and technology development.

The term "engineering" is used in this book for both engineering work and the engineering technician profession. The book focuses on general issues for engineering but examples for different engineering disciplines are mentioned in different contexts. In addition various types of organizations for technical education are collectively referred to as "technical university".

Social aspects of technical education challenges can be identified and categorised from three perspectives as follows:

* The first point of view addresses the factors necessary for the success of engineering work in the future and the associated potential restriction. A substantial portion of the analysis presented in this book (Part I) is allotted to identification of engineers' opinions regarding the future and the different implicit and explicit social values present in conversations and statements within the engineering community. Information used in this analysis is collected through literature survey and desk research focusing on the development of the engineering profession as well as a survey through structured interviews to investigate the required engineering capabilities. The survey methodology is described in Chapter 4.

* From the view point of technology development in the 21st century requirements for technical education are discussed in a broader context of engineering work (Part II). Different roles of technology in society are discussed in Chapter 5. A robust concept of a socio-technical system is developed demonstrating different possible tasks for engineers during the technology development process (see Chapter 6). Not all of such tasks are yet basic engineering works (see Chapter 7 on reflexive knowledge). The developed concept of a socio-technical system is also used to point out uncertainties and challenges related to technology development which influence the future of the engineering profession.

- From the view point of technical universities the challenges regarding the organization of research, education and administration at technical universities are identified. This analysis is based on the understandings from the previous two perspectives regarding the success factors influencing the engineering work in the future as well as challenges to technology development and the design of new strategies for the universities. Results of this analysis together with the conclusions of Parts I and II are used for identification of crucial questions for technical education. Chapter 8 emphasizes the social responsibilities and future impacts of engineering work in sustainable development. Chapter 9 includes objectives for technical universities.

The findings from these analyses are integrated in Chapter 10 in preliminary recommendations regarding the pathway for overcoming the identified challenges to technical education in the 21st century.

Results of research are summarized in Chapter 11.

These recommendations are made by considering the preconditions of the socio-political dimension for sustainable development based inter alia on long-term strategies, democracy, fairness and participation as principles (see Albert et al. 2001).

The recommendations introduced in Chapter 11 are "preliminary recommendations" since they are based on the specific analysis presented in this book which the author considers as a work still in progress. Dougherty (2003) has described the problem of limitations due to the individual assumptions in the analysis of issues and the corresponding hypothesis generation. These recommendations therefore should be corrected and enhanced from many different perspectives before they can form viable hypotheses.

An issue that the author encountered during this analysis was that public debates and scientific references are available primarily on technical innovations and not on issues related to the engineering profession itself. From a public point of view the link between the engineering profession including technological development and technical education is an internal affair for the engineering profession. Part II therefore starts with a new view of a socio-technical system that considers a number of interactions between technology development and engineering work. This concept is used to provide an unprecedented view of the functional issues, structural elements and uncertainties of technological development and their links to both the engineering profession and technical education.

2.1 The role of examples and statements in making the recommendations

A principal element for the adaptation of first positions and making the recommendations was to search for examples and statements on different social values that influence technology development, the engineering profession and technical education. These examples are not analysed to check any hypotheses, but instead are used to identify diversity of needs and to show that there are different ideas and opinions regarding the role of engineering work and the future of technical education. Recommendations (listed in Chapter 10) on the pathway towards the new technical education needed to take into account all different opinions and points of views on the subject.

The purpose of using original texts was to show different terminologies by different relevant disciplines and people involved in technology development and engineering work. The term "people involved" is used for describing individuals who influence or are influenced by technical education. "Stakeholders" are therefore not only the interested people but also those who can influence technical education.

The author emphasises that she has used certain statements by different scientists to formulate the first ideas followed by making new statements and recommendations. The reader should be aware that the author has presented only one possible interpretation of these statements and references. The cited theories described by the social and human scientists such as Feenberg, Luhmann, Habermas, etc. are much more complex than the parts cited in this book.

Examples and statements are investigated through a literature search as well as in a questionnaire survey.

The survey was an element to identify needs for the engineering of the future using the language of engineers from different disciplines and to compare the success factors which are mentioned by individual engineers for different contexts. For this comparison it was necessary to study not only texts related to companies, technical universities and relevant policies but also to conduct an explorative survey on opinions of individual engineers at a micro level. The method of the survey and the questionnaire are described in Chapter 4 that also includes the results of the survey. The importance of original citations for terminology is presented in the next section.

2.1.1 The "precautionary principle" terminology

In order to show the role of original citations and examples in this analysis, this section presents an example on "precautionary principle". This book contains terms such as "precautionary principle" that are based on a number of different issues. Such terms are controversial with different interpretations. The following definition of the precautionary principle refers to "uncertainty", "risk" and "precaution":

> "Precautionary principle helps policy-makers and politicians, in circumstances in which waiting for very strong evidence of harm before taking precautionary action, may seriously compromise public health or the environment, or both." (EEA 2002)

Furthermore the terms "precautionary actions" and "prevention" play an important role in the definitions. A major challenge in the decision-making process is the uncertainties about the impacts of complex systems. The example of ozone depletion in Box 1 shows that preventive measures for the environment and health are needed even in the absence of scientific certainty, when the potential risks are at unacceptable levels for human beings. The next definition adds the term "economy".

> "Waiting for incontrovertible scientific evidence of harm before preventive action is taken can increase the risk of costly mistakes that can cause serious and irreversible harm to ecosystem and human health and well being, and the economy"[9]

The example on ozone depletion also addresses the term "ignorance":

Box 1

Example of stratospheric ozone depletion

"Man-made ozone depletion is caused by chlorine and bromine, but not all chlorine- and bromine-containing compounds are harmful to the ozone layer. A large number of compounds react with other gases in the troposphere or dissolve into rain droplets and do not reach the stratosphere. The longer the atmospheric lifetime of a compound, the more it can enter the stratosphere. The chlorine and bromine species that cause depletion of the ozone layer are CFCs, carbon tetrachloride, methyl chloroform, HCFCs and halons, all of which are entirely of anthropogenic origin. The ozone layer can also be depleted by methyl chloride and methyl bromide.

Emissions of CFC-11 and CFC-12 began to fall in 1974, following reductions in their use as propellants in aerosol spray cans, resulting from concerns triggered by publications in the early 1970s suggested that CFCs could deplete the ozone layer. Emissions rose again in the early 1980s, mainly from non-aerosol uses such as foam blowing, refrigeration and air-conditioning, and fell after 1987 in response to the Montreal Protocol.

The limitations imposed on the production of CFCs triggered the use of HCFCs and hydro-fluorocarbons (HFCs) as replacement compounds. HCFCs contain chlorine and can affect the ozone layer, but much less than the CFCs they replace. HFCs do not destroy ozone (but are greenhouse gases and belong to the basket of greenhouse gases agreed upon in the UNFCCC Kyoto Protocol)."

---

9   Http://www.biotech-info.net/final_statement.html, accessed 13 May 2008.

Methyl Bromide is another gas that can deplete stratospheric ozone. Global emissions and sinks of methyl bromide are not well understood. Anthropogenic emissions come from agricultural usage (mainly soil fumigation, 31% of total emissions), biomass burning (22%) and gasoline additives (7%), with minor contribution from sources such as the fumigation of buildings and containers (3%) and industry (2%). The largest natural source of Methyl Bromide is the oceans (35%), but they also act as a large sink, making their overall role in the global budget of methyl bromide difficult to assess (SORG, 1996). Other sinks involve atmospheric oxidation and soil uptake.

Halons, CFCs, CCl4, CH3CCl3 and HBFCs have already been phased-out of production due to the Montreal Protocol in developed countries."

(European Environment Agency 1998)

Other sources of ozone depletion are emissions of nitrogen oxides, water vapour, and sulphur dioxide from aircraft exhausts.

The example in Box 1 shows the need for the precautionary principle and an early evaluation of new products impacts, when there is a fear that emerging technologies will cause more problems than they solve.

Sanderson (2002) discusses the scientific application of precautionary principle in a sustainable development as follows:

"Since the Rio meeting on sustainable development in 1992 the precautionary principle has been written into many different national and international treaties and conventions. The precautionary principle/approach can be seen as a government's tangible commitment to the importance of social values such as health, safety, the environment and natural resources conservation. Principle 15 of the 1992 Rio Declaration on environment and development states that: ... lack of full scientific certainty shall not be used as a reason for postponing cost-effective measures to prevent environmental degradation.

... The reason for implementation of a precautionary principle is that while scientific information is still inconclusive, decisions will have to be made to meet society's expectations about living standards and to address risks. The scientific process is, and should be, almost always [characterised] by uncertainty and debate, which is consistent with Sir Karl Popper's scientific falsification theory (1968). The challenge to ... scientists and to society is to determine what is sufficient scientific certainty to implement a precautionary approach and furthermore, how to achieve a scientific application of precautionary approaches within research?"

Sanderson (2002) claims the typical case for using the theory of post-normal science when

"... facts are uncertain, values in dispute, stakes high and decisions urgent. ... It is for this reason that the quality of the procedures of inquiry depends on an 'extended community peer review', including all with a concern for resolving the issue. It could be said that the theory of post-normal science is the essential foundation for the [realisation] in practice of the precautionary principle in science related policy issues in light of epistemic uncertainties."

(Funtowicz/Ravetz 2001)

He introduces a list of typical topics for discussions among scientific communities and policy makers towards the precautionary approach:

1. What is known?
2. The certainty with which it is known.
3. What is not known?
4. What is suspected?
5. The limits of the science.
6. Probable outcomes of different policy options.
7. Key areas where new information is needed.
8. Recommended mechanisms for obtaining high-priority information (see Sanderson et al. 2002).

The European Environmental Agency (EEA) released twelve lessons from early warnings based on precautionary approaches from 1896-2000. These "lessons" are as follows:

1. Acknowledge and respond to ignorance, as well as uncertainty and risk, in technology appraisal and public policy-making.
2. Provide adequate long-term environmental and health monitoring and research into early warnings.
3. Identify and work to reduce blind spots and gaps in scientific knowledge.
4. Identify and reduce interdisciplinary obstacles to learning.
5. Ensure that real world conditions are adequately accounted for in regulatory appraisal.
6. Systematically [scrutinise] the claimed justifications and benefits alongside the potential risks.
7. Evaluate a range of alternative options for meeting needs alongside the option under appraisal, and promote more robust, diverse and adaptable technologies so as to [minimise] the costs of surprises and [maximise] the benefits of innovation.
8. Ensure use of "lay" and local knowledge, as well as relevant specialist expertise in the appraisal.
9. Take full account of the assumptions and values of different social groups.
10. Maintain regulatory independence from interested parties while retaining an inclusive approach to information and opinion gathering.
11. Identify and reduce institutional obstacles to learning and action.
12. Avoid "paralysis by analysis" by acting to reduce potential harm when there are reasonable grounds for concern. (EEA 2002)

The texts quoted in this section show the diverse terms that are used to describe "precautionary principle". Such definitions themselves have different possible interpretations that need to be clarified. In the case of precautionary principle the author has shown texts from Sanderson (2002) and the report of EEA (2002) on "Late lessons from early warning" that include different citations of researchers from social and natural sciences to show an example of different scope and depth of existing interpretations of such terms. The cited text therefore should help readers to compare their own interpretations with those from authors in other disciplines.

Part I

Engineering

3. Development of engineering challenges

What was the role of engineers in our social and cultural development?

The future challenges for engineers can be predicted by understanding past situations, analysing the present issues and estimating future needs. This chapter starts with a short history of engineering in section 3.1 and continues with the analysis of some of the 20th century issues in section 3.2. Future engineering trends based on literature research and interviews are described in Chapter 4.

3.1 Roots of engineering as a career

The historical role of engineers in society is briefly addressed here. In this section the author presents some of the historical achievements by engineers and a short history of engineering challenges in relation to policy and society.

The historical contents below are quoted from direct citations and author's translations (Rae/Volti 1999; Kaiser/König 2006; Canel et al. 2000; Jischa 2004; Wulf 1998) as well as citations from the internet sites of engineering associations, technical universities, technical museums, etc. The quoted citations show the original text by their corresponding authors and organizations. This is necessary since the references are from different disciplines and cultures and the author intends to show the diversity of the languages among these disciplines.

Historical evidence for engineering work starts perhaps with the works of Imhotep the chief minister of the Egyptian Pharaoh Zoser (c. 2750 B.C.) including the construction of buildings and water supply canals.

There are historical documents dating back to around 400 B.C. showing the original use of mechanics and its prominent role in the eastern and western parts of the world. The designers, researchers, craftsmen and organizers who built palaces, temples, harbours, etc. were called "architekton" in Greece and "architectus" in Rome. The history of technical solutions shows most achievements as being the ideas of political and religious leaders themselves. Only a few engineers such as Hero (or Heron) of Alexandria (c. 10-70 A.C.) are famous for their own magical deeds. Hero's famous fountain is well known today, and his "miracles for temples" such as the self-opening door or thunderous sounds are documented in his book Automata. These constructions were used to impress the believers.

> "Hero of Alexandria was a celebrated mathematician, physicist and engineer who lived in the 1st century A.C. He set up and directed the Higher Technical School of Alexandria, which he developed into a genuine Polytechnic. Often referred to as the "encyclopaedist", he built on and developed the theoretical

and practical work of Ctesibius, whose famous constructions place him among the greatest figures in mechanical engineering in the ancient world."[10]

Pioneering role of China is destroyed by opium war:

In the aftermath of the ancient world, technical development took a different path in each part of the world. In the East the development of hydraulic engineering, measurement instruments, war techniques, shipping, and paper production are examples of such developments which continued in China even into the modern times of the 19th century. The opium war in the 19th century destroyed the long tradition of the pioneering role of the Chinese in technical innovations. The Chinese art which has been widely distributed in society has however survived to this day.

The relationship of engineering with culture and language:

Technical development was at a high level in Persian culture until about 700 A.D. Persia's loss of independence through Arab invasion and the ban on the Persian language also affected the technical development and engineering achievements.

Hill gives many examples of early engineering design in the Middle East in his book "A History of Engineering in Classical and Medieval Times" (1996).

In the 12th century the European technology and engineering for civil applications began to develop rapidly. Development of cities increased the demand for buildings, water supply systems, better medicine and defence technology. "Ingeniato" is considered the root for the word "engineering" for the first time in texts from the 12th century A.D., mainly used in the context of military applications.

Difficulties of communication of ideas as an obstacle to engineering success:

History shows that the difficulties in communication of ideas have posed many challenges for engineering activities. Language and art were the two main drivers for the generation, communication and distribution of innovative technical designs. Lack of communication of ideas among scholars and future generations retarded the development of technical solutions. Examples are the loss of the Persian engineering know-how as a result of the loss of the language for communication between masters and craftsmen between 700 and 1500 A.D. resulting from the restrictions of using Arabic for scientific and official documents. A similar situation occurred in Europe due to the difficulty of popular access to knowledge caused by among other things, the rigid use of Latin for scientific texts and technical documents until 1600 A.D. The failure to communicate knowledge slowed down the distribution and accumulation of new knowledge and hindered technical improvements.

---

10 (Technology Museum of Thessaloniki 2001).

Lessons from history as positive arguments for a broader engineering knowledge for individuals and a broader technical understanding for society

The importance of the communication of knowledge can be shown in the decrease in the technological development of occupied countries due to the destruction of libraries in wars. Other important factors were political situations and the social potential for technical innovations. The history of China, Persia, Rome and Greece shows one main risk of an isolated and elite technical development in ancient and feudalist regimes: If only a small group in society is aware of engineering knowledge, there is a high risk of losing the knowledge during crises such as war or population losses during natural disasters, epidemics, etc.

Individual engineers needed mobility and presentation art:

The language barrier for people without the knowledge of Latin was overcome in Europe at first by a combination of art, science, practical knowledge and organization intelligence, as well as by the mobility of artists such as Leonardo da Vinci during the Renaissance period. The drafts for buildings and the three dimensional models of constructions which were used around 1500 A.C. in Europe served two purposes. They were means to communicate plans to sponsors, princes and lords of castles as well as being a documentation tool allowing the next generation to continue and even improve the designs. The construction of the Basilica di Santa Maria del Fiore in Florence, Italy, is an example of such continuous improvements. The cathedral is noted for its distinctive dome. The first stone of the building was laid on September 9, 1296. The nave was finished by 1380 and by 1418 only the dome was left uncompleted. The 42 meter (137 ft) wide space originally had a wooden dome, built by Arnolfo di Cambio. Construction of a stone cupola over the chancel posed many technical problems. A brick model for the dome existed from 1367. In 1420 Filippo Brunelleschi (1377-1446) won a competition for the construction of the dome. He drew his inspiration from the double-walled cupola of the Pantheon in Rome. He constructed a wooden and brick model with the help of Donatello and Nanni di Banco (on display in the Museum Opera del Duomo). His model served as a guide for the craftsmen, but was intentionally incomplete, so as to ensure his control over the construction.[11]

Combination of art, science, tacit knowledge, organization and economic tools:

Brunelleschi was an example of a generic engineer who used art, science, practical knowledge, organization and economic tools to implement his plan. He designed machines to transport heavy materials to the top of the building; he made contracts to transport materials by ship close to the building site; he used new

---

11 Http://www.mega.it/eng/egui/monu/bdd.htm (The Florence Art Guide), accessed 13 May 2008.

techniques in scaling up his 3D models step by step. In a period of about 20 years, which was a short time in that epoch, he was able to finish the dome.

The combination of science, practical knowledge and art reached a new level of excellence with Leonardo da Vinci (1452-1519) (see also Jischa 2004)[12].

An individual interdisciplinary learning program:

Leonardo da Vinci originally learned mathematics in an abacus[13] school for traders. He studied Latin and mechanics on his own and, parallel to his paintings, developed a rich collection of innovative ideas, designed drafts of flying machines, water supply systems, textile making machines, clocks, diving-suits, weapons, etc. His work strongly addressed an elite and military aspect of technical design[14].

Protection of the technical knowledge was a challenge for economic success:

Later in the 16th century it became important for princes and lords of castles to protect the knowledge and technical innovations within their territories for example ovens. The system of patents was developed during this time in north Italy and Austria for small ovens. Patents were intended to improve the distribution of news on technical innovations and protect the exclusive economic profit from engineering achievements for inventors and investors. Another means for distribution of technical innovation was books with schematics or technical drafts. Inventors described the principle of mechanical instruments in these books, but in order to prevent their innovations being copied, they did not show the manufacturing process.

Trade off between distribution of applications and protection of knowledge:

During this period of the spread of technical applications, different regions started to set limits for certain practices. In Amsterdam for example, it was forbidden to use windmills for stone breaking in the city.

More systematic education for more craftsmen:

Craftsmen in Europe had to use books, models, drafts, standards, patents, etc. to communicate with investors and decision-makers, and the demand for skilled craftsmen increased. This laid the ground for the plans for technical schools in a similar and comparable form in Europe in the 16th century. Technical education was brought into scientific academies in the 17th century and evolved continuously

---

12 The importance of experience was denied up to this time due to the opinion of Aristotle and Plato whose ideas had shaped the science world since the Renaissance. They did not believe in practical experience as a true knowledge.

13 Glossary.

14 Military engineering is one of the main and oldest engineering fields. Since this field is not in focus of this analysis, it will not be discussed here.

into separated technical schools with different engineering disciplines in the 18th century. These universities ensured the availability of engineers for the major projects of the late 17th century following the industrial revolution. During this time there was a high demand for more engineers with different knowledge levels to work in the hierarchical organization of companies. One of the engineers who proposed the concept of technical education was Christopher Polhelm.

Box 3

Lessons from history

Education – Laboratorium Mechanicum[15]

Christopher Polhem (1661-1751-Sweden) was an inventor and entrepreneur who thought that a technical laboratory ought to be an important part of the education of future engineers. In 1697 he set up his Laboratorium Mechanicum with the purpose of teaching research and demonstrating to visitors all that technology and mechanics could do. Tradesmen would work on producing models, while Polhem as the principal performed research and instructed the students with the help of the models.

The Mechanical Alphabet

One aspect of Polhem's teaching method was to illustrate all the mechanical components that a designer needed to know. He called his collection of wooden models the "Mechanical Alphabet", which showed simple principles for the conversion of motion and was used in his teachings. Today, the remains of the chamber of models can be found in the National Museum of Science and Technology's collections (Tekniska Museet) in Stockholm.

Self-taught engineers in England developed the civil engineering profession:

The idea of technical education and a fixed school program was not initially accepted in all European countries. Until the 19th century Great Britain had a system of practical engineering learning. Scientific knowledge was easily available for traders and craftsmen and self-taught young engineers could learn the profession by doing. James Watt (1736-1819) is a good example of the successful strategy of this system in England. He was a craft worker or technician in the field of scientific measuring instruments. He was interested in scientific issues and gradually evolved into an engineer during his cooperation with the businessman Matthew Boulton (1728-1809) on the steam engine (Cornish Mining World Heritage 2007). The improvement of the steam engine happened during these years and changed the technology of power generation for the textile industry, transport and many other applications. During the industrial revolution in England it was not really necessary to receive a standardized technical education in order to be able to perform as an engineer. English engineers who learned the profession by themselves such as Thomas Telford (1757-1834) and Thomas

---

15   (The National Museum of Science and Technology in Stockholm 2006).

Tredgold (1788-1829) contributed to the development of the Civil Engineering profession and its separation from military engineering. Tredgold described the profession as: "… being the art of directing the great sources of power in nature for the use and convenience of man" (Roman 2003).

During the industrial revolution, as the demand for natural resources and working power increased, the influence of economic power increased on the engineering profession alongside the existing influence of political power. Engineering improved more to become the profession that used knowledge for industrial development and to provide economic benefits to the society (or a special class of society). The engineering profession merged with the industry and became rather isolated from the systematic scientific feedbacks. Engineers worked under the instruction of a board of directors who had their own ideas on design. I.K. Brunel (1806-1859) who designed and managed the construction of the Great Western railway in Britain played an important role in the emancipation of civil engineers in the society from these boards of directors. He demanded more confidence in engineers.[16]

Lack of standardized technical knowledge was a challenge to the profession: After the deaths of the English engineers and inventors Brunel, Robert Stephenson (1781-1848)[17] and Joseph Locke (1805-1860)[18] within a period of ten years the image of the engineering profession started to fade in England.[19] Due to the lack of a standardised technical education in England and the less science-friendly industry at that time in comparison to other western countries, the engineering profession could not remain vital without excellent engineers as leaders of the engineering communities. Local and private initiatives and informal education were not enough for England to compete with France and Germany. The economic changes in England and the decreasing role of engineers are considered as interwoven phenomena. The elite kept their cultural statues and protected the language and culture against the French and German languages and the majority of engineers transferred to the working class. The middle class did not choose technical education for their children and emphasized elite social behaviour or economic education. The contact between the industry and innovative self-taught individuals with an interest in engineering profession as a career was therefore broken.

Decline of the engineering profession image in England:

In contrast to the decline of the status of engineering in England, American engineering started its real development around 1870 and left other western countries

---

16   See "obituary for Isambard Kingdom Brunel appeared in The Times on September 19, 1859" (The Times 2001).
17   (Bellis 2007).
18   (Spartacus educational 1999).
19   (Rolt 1958).

behind in one century. At the beginning of the 19th century America still needed European engineers for the construction of bridges, canals, etc. Most of them such as Brunel went back to Europe for that purpose. During this period of time a number of American engineers studied in Europe for some years.

The relationship between industry and engineering:

One of the reasons for the rapid development of engineering in America was perhaps the difference between the European and American industries in the 19th century. The structure of English industry was characterized by small family enterprises growing into large bureaucratic companies. In contrast, industry in America during that time was flexible and more interesting to engineers. In America the majority of engineers enjoyed a practical education while a small group were taught in technical schools with a strong theoretical basis. The system was a mix from England, France and Germany. American engineers could take on management positions and they enjoyed a better image than engineers in England.

A balance of scientific and mathematical fundamentals:

The second half of the 19th century saw a turning point in teaching philosophy at some of the American colleges. In 1885, the Sibley College of Mechanic Arts and Mechanical Engineering decided to set up the mechanical engineering program according to a new educational model based on a balance of scientific and mathematical fundamentals, engineering science, and practical shop-floor experience. A European collection of mechanical instruments was the cornerstone of practical teaching (Moon 2005).

Alongside the scientific fundamentals of mechanical engineering, management knowledge was integrated into construction engineering in America. The example of John A. Roebling shows that not only mechanical engineering but also civil engineering was influenced by the European experience.

> "John Augustus Roebling (born 1806 in Mühlhausen/Germany – July 22, 1869) was famous for his wire rope suspension bridge designs, in particular, the design of the Brooklyn Bridge. He graduated in Germany in 1826 with a degree in civil engineering. Additionally, Roebling studied under the famous German philosopher Georg Hegel. In 1831, Roebling immigrated to the United States. His first engineering work in America was devoted to improving river navigation and canal building around 1840. In 1841, he began producing wire rope at his workshop in Saxonburg. He had been fascinated by the idea of suspension bridges since his college days, and wrote his graduation thesis on the subject. In 1867, Roebling started design work on the Brooklyn Bridge spanning the East River in New York. He died from an infection before the bridge was finished. Roebling's son Washington continued his work on the Brooklyn Bridge."[20]

---

20 Http://www.inventionfactory.com/history/RHAgen/jarbio.html, accessed 13 May 2008.

"After Washington Roebling died his wife Emily Warren Roebling (1843-1903) continued the project. Trained in mathematics she learned to speak the language of engineers, made daily on-site inspections, dealt with contractors and material suppliers, handled the technical correspondence, and negotiated the political fictions that inevitably arose in grand public projects."

(Oldenziel 2000, P. 15)

Anonymous female engineers:

In May 1883 it was Emily Roebling who was the first person to ride with President Chester Arthur across the great bridge.[21] During this period most female engineers worked anonymously under the name of their fathers, brothers or husbands, even if they had a formal engineering education. Graduation in a formal manner was also difficult for women. They could take part in mathematics exams and laboratory experiments, but were excluded from important workshops and practical training at building sites necessary for graduation. Although the entrance of women into military engineering was easier in times of war, there being a lack of men, they had to work under masculine managerial command. Female engineers were therefore excluded from the professional advantages of individual engineers. Until the 1950s, female engineers in most of the European countries, America and Russia attended courses in architecture, chemistry, and new fields not yet dominated by the "male culture of technology". In architecture they often operated as engineering assistants, and chemical engineering was connected with a laboratory career. Centralized big industry was at that time a male-dominated engineering environment with female engineers occupying the role of assistants only. Female engineers achieved more success later in specialized new fields such as information technology, biochemistry or electronic engineering working for private industry. After a difficult start for female engineers in the nineteenth century, the twentieth century put female engineers with excellent qualifications at the lower organizational levels. Engineering was segregated into drafting and laboratory work for women as "engineering assistants", versus management and jobs at production sites for male "engineers". In 1948 the Women's Bureau reported that most women trained in the field of chemical engineering were employed as chemists. Choosing the engineering occupation was not a task for women thanks to male-dominated rules and traditions at universities, in industry and in professional societies.

The situation was not much better for the British female engineers. The post-war industrial world, especially after the First World War, was concerned about the occupation of engineering positions by women and considered the work of female engineers as less productive. Women in engineering in Sweden faced social pressure and the admission of female students into universities was possible

---

21 Http://www.asce.org/history/bio_roebling_e.html, accessed 13 May 2008.

only through individual selection process until 1921. Subsequently, formal new rules allowed for the education of female engineers, but the informal culture was still a public world dominated by men. Women performed engineering tasks under the supervision of male engineers. Gradually divisions formed among draft work of less prestigious activities at laboratories, production sites and outside work of elites. Women worked in those new low-prestige fields in laboratories and as draft workers in electronics and chemistry fields. The outside work remained the domain of male engineers.

In Russia engineering existed as a profession at the turn of the eighteenth century. The first high-level school in Saint Petersburg was established in 1809/1810 using French education models and the French language, which was the scientific language at that time. Women could attend engineering schools between 1835 and 1863 but they were still outsiders. Many female engineers were politically active and supported radical political movements. For this reason a restriction on women studying was introduced during the revolution in 1863. Many Russian women students continued their engineering education abroad and some developed the system of home universities called *Universities on the Wind* which were very flexible and mobile and attracted professors such as D. Mendeleev. In 1878 the Higher Courses for Women started their work as an institution similar to a university. The graduates could teach at high schools and the diplomas were made equivalent to a university degree in 1913. In 1915 The Russian government finally recognised women's professional skills as engineers. By 1916 female engineers had already participated in construction projects for railways, bridges, plants, power stations and ports.

In France female engineers started their training in women's polytechnic school in 1925 in a program at the University of Science in Paris. French male engineers occupied an elite position as high ranking civil servants. Striving for engineering education in existing male only institutions was therefore more difficult for women. However in the 1920s female engineering students studied at the Ecole Centrale to get engineering jobs in industry. The civil works for state and military engineering were reserved for men. Marie-Louise Paris (1889-1969) who was a promoter of engineering for women in France recognized the engineering opportunities in the fields of laboratories and drafting that were considered more suitable for women. As industry expected another war in 1938 and the need for more engineers increased, the Women's Polytechnic school was listed as one of the schools authorized to award the title of engineer. In 1956 the school after ten years of a nomadic existence was moved to a building in Sceaux near Paris which was bought by M.L. Paris, the president of the CNAM having forced the school to move out of the university building. Industrialists who needed specialized professionals as well as bright scientists and engineers supported the idea of courses in computer science, solid-state physics, and aerospace at schools.

"Maurice Berthaume, an eminent professor of aerodynamics, who had a management position at the major French aerospace corporation, l'Aerospatiale, even left his job to run Paris' school of Paris after her death in 1969."

(Canel 2000, P. 142)

The school used industrial grants, M.L. Paris' own capital and support from the parents of female students. Instead of pushing women into the established world of men's engineering, Paris managed to achieve a high standard of courses and laboratory work for students and offered new job opportunities by finding new niches for her students in a period of deep economic crisis.

In Germany the 1930s were a time for immigration for many scientists and engineers. A number of female engineers moved to other countries. A group of female students was not allowed to continue their studies at their university. The small number of female scientists and engineers after World War II could not be increased immediately after the war. In the early 20th century in Germany engineering profession was merged into industrial institutions that were partly public and partly private. The public institutions were not as flexible as the private ones due to restrictive regulations imposed upon female engineers. Private institutions had better access to female scientists and engineers, paying them the same amount of money as their male counterparts in similar positions and in some cases even promoting women engineers to be head of a laboratory or department. The biography of female engineers during the Second World War shows that women often worked for the armed forces. Some of them were forced from their jobs when they were married. After World War II women were moved from their positions in Germany. Between 1960 and 1970 the number of female students in engineering sciences rose in West Germany and the GDR. Chemical engineering was one of the engineering fields with a high share of female students. Some European female engineers were able to continue their careers in scientific research in the United States and other countries. Irmgard Flügge-Lotz (born in 1903 in Hameln in western Germany and died in 1974 in Stanford, USA) became the first female professor in engineering at Stanford University in 1961.

Some further examples of engineering development in USA:

The engineering profession changed especially in America between 1880 and 1945 from an individual form for the private sector with innovators such as Thomas Edison to a mass profession for large companies. Research and development at public laboratories also expanded for military engineering purposes. In the 20th century more specialisations emerged in engineering, and mechanical and civil engineering became two separate fields. After the First World War these two fields together with chemical and electrical engineering accounted for the main engineering fields. Large-scale industrial research needed the cooperation of thousands of engineers with special responsibilities in a hierarchical structure.

The idea of research for systematic technical development led to new requirements for the engineering profession

Engineers after the Second World War needed to learn to work in complex industrial plants. Individual management techniques and talents were not enough for their careers. The scientific and business basics of engineering were added to the practical knowledge and led to a change in technical educations, after 1945. Practical work was planned for engineering technicians, and engineers moved to a variety of careers. The merger of engineering and management in the multi-level management structures of large industrial organizations was the challenge of the engineering profession in the twentieth century.

The technocracy movement at the beginning of the twentieth century supported by Howard Scott proposed the idea of scientific management and technocratic governance.

> "The Technocracy Movement began in 1918-1919 when Howard Scott brought together a group of economists, engineers and scientists for the purpose of creating a research organization. This organization was known as the Technical Alliance throughout the 1920s and the early 1930s. In 1933 the group was incorporated according to New York State laws, and officially became Technocracy Inc., a non-profit, non-political, non-sectarian membership organization."

> (Hill 2005, see also Akin 1977)

Technocracy put technology and the efficient use of energy at the centre of the decision-making process. One phenomenon related to this philosophy was the focus of technical education on natural science and an emphasis on the economics of technical innovations for economic growth in the 20th century. Today we can observe a new kind of technocracy movement in the form of eco-efficiency, CO_2-trading systems and other strategies that regard technical innovations as the key driver of economic growth with a win-win effect for environmental protection. The benefits and limits of the eco-efficiency concept will be reviewed in section 4.1.

Despite the scepticism within a majority of technical universities who favoured the specialisation of engineers, inter- and trans-disciplinary engineering education was developed in a number of countries. In the 1960s an optional program in urban planning was established in France based on multi-disciplinary methods that helped students to work with users. The integration of user needs and the study of interactions between humans and machines was one step towards admitting the influence of society on the development of technical solutions. User integration is today an important part of computer engineering.

Engineering is today a broad field with very different specifications and functions. There is no such thing as a general engineering practice thanks to the specialization trend in the 19th and 20th centuries. Engineers have different roles in society. In a company they can be staff, CEO, consultants, etc. They also work as lecturers and policy-makers. In each of these positions they have different roles and responsibilities. There is however limited information available on the comparison of engineers' behaviour and performance in these roles.

3.2 A short look at the present engineering challenges

Is engineering today only the application of scientific and mathematical principles to the design, manufacture, and operation of efficient and economical structures, machines, processes, and systems?

Is it a skilful manoeuvring of direction?

What are the present engineering challenges from the perspective of the engineers' themselves?

In this section the author briefly presents a discussion for better understanding of the present engineering challenges. Addressing these challenges is an important step towards transformation of technical education of the 21st century.

Discussions on an open Internet forum among engineers have identified some important challenges for the engineering profession as follows:

- the need for new problem-solving skills;
- acceleration of work load;
- lack of appropriate quality management[22];
- risks and opportunities associated with the use of information technology in engineering work;
- engineers' responsibility during different design phases, in a fragmented design process;
- interdisciplinary and team work with non-engineers, etc. (see Appendix A).

The forum discussion took place during the months of January and February in 2006. The majority of the participants in the discussion indicated that the diffusion of information technologies has substantially changed the engineering design process which happens in a visual system with the involvement of many contributors. Some of the participants saw the division and fragmentation of engineering work as a problem and said that it is difficult for individual designers to have an overview of the entire design project. Other participants described the benefits of information technology and the Internet for their designs and businesses. There were positive statements regarding the connectivity between local and international activities created by the Internet. The forum discussions summary is presented in Appendix A.

---

22 Glossary.

The results of a literature survey showed a number of more underlying challenges for engineering work as follows:

- local engagement and global responsibility;
- responsibility for health risks and industrial protection;
- women in engineering;
- the tension between long-term goals and short-term strategies;
- the tension between engineering and science;
- obstacles to research and development;
- public image of the engineering profession.

In the next sections short descriptions of these challenges are presented and the role of technical education in dealing with them is discussed

3.2.1 The challenge of engineers' responsibility at local and global levels

Are engineers and engineering communities independent enough to push industries and governments for higher technical standards at the global level?

The requirements of the precautionary principle and fairness at local and global levels pose one of the main challenges for the future engineering profession, namely the definition of its role in both the local community and the global society. Engineers could either take on the responsibility for understanding the needs and problems of their society at both local and global levels or they could work in isolation within the framework of the rules set by the industry or regulatory organisations. The literature search conducted in this study showed that engineers often perceive their responsibilities within the framework of regulations and norms developed for industrial or research organisations. During this research the author was only able to find a few examples of autonomous commitment by the engineering profession to the precautionary principles and fairness at local and global levels. There are a number of engineering societies and activities focusing on global responsibility. For example, Engineers Against Poverty (EAP) is such an international engineering society.[23] Nevertheless the engineers' roles are often unclear in communal activities of social projects or international projects administered by the United Nation's units and programs, for example. Nor

---

23 "EAP is a specialist international development NGO working in the field of engineering and international development. It was established in 1998 by the UK's leading professional Engineering Institutions, the Royal Academy of Engineering and the Department for International Development (DFID). It is incorporated as a private company limited by guarantee and registered as a charity. EAP is working to become the leading agency allied to the engineering profession, working with industry, government and civil society to fight poverty and injustice and to promote sustainable development." http://www.engineersagainstpoverty.org/index.asp?SectionID=2, accessed 13 May 2008.

it is clear whether the engineering profession can generate initiatives for more responsibility for industrial production at local and global levels. Technical education in the 21st century should generate the awareness and support teaching the skills that engineers need for taking a more pro-active and responsible roles related to the environment at local and global levels.

This is a new and different way to implement a technocracy strategy. For example individual engineers should be more responsible for and be aware of precautionary principle and fairness. Traditional technocracy focuses more on the need for organizing engineering communities and the need for more efficient technical solutions for improved productivity. One similarity between the two approaches is the requirement for an active role to be played by engineers in improving the norms and standards for both technical excellence and environmental protection. In fact standards and norms are not a new instrument for change. They are often a signal from the state of the art and require a rather long time for their development and implementation. Engineers might be faced with situations on a daily basis that affect their local or global societies yet they are not regulated through any norms or standards. In some cases engineers are not aware of the potential impacts while in other cases they can estimate the high potential risks or are aware of uncertainties about the impacts but cannot change the rules of practice. CFC production and release into the atmosphere and the corresponding negative impact on the ozone layer is an example of such problems. In the first few years the impacts of innovative products manufacturing were not known, and after scientists began to point out the impacts on the depletion of the ozone layer, it took a long time for a ban to be imposed as a result of the long and complex process of control norms, standards and conventions development and implementation.

The issue of CFCs' undesired impacts could perhaps have been avoided through an intensive research and analysis of such potential impacts or if engineers' attention had been drawn to the issue from the beginning or if they had a better communication with CFC-sceptical natural scientists and researchers. The acceptance and application of the precautionary principles by the actors involved, such as engineers, can lead to more precaution in the absence of standards and regulations reducing the potential risks of negative impacts in the future. A detailed discussion on the role of norms and standards at local and global levels in directing the engineering profession in the future could provide more insights into the content and composition of the courses for the engineering students. This discussion is however beyond the scope of this book.

We should also be aware that global responsibility is completely different from global engagement. Engineering work is a global activity in many situations today. The global engagement through engineering work has caused serious problems since the technical design of international projects have not always been compatible with the specific local climate, culture, social values, resources or economic conditions. Negative impacts of technical solutions can also arise in the case of isolated local solutions that neglect global responsibility. The impacts of engineers' designs should therefore be evaluated in light of both local and global environments. The design of energy efficient buildings from locally-available environmentally-friendly materials, efficient and environmentally-friendly water supply systems, and the design of low emission transportation systems taking into account the needs of the different groups in the local community, and the implementation of integrative measures in manufacturing sites to prevent air emissions are some of the examples that combine the local commitment with the global responsibility.

The challenges facing such engineering work are presently the costs and the necessary communication skills for the technology developers to interact well with their local communities as well as the need for the knowledge and financial resources for addressing global issues. Finding a solution for a balance between local commitment and global responsibility therefore remains a considerable challenge for the engineers of the future.

Technical education should in this case provide the future engineers with the awareness and knowledge to understand and address both local and global issues and to take individual responsibility to bring out and discuss the limits of the available standards. They should also learn the necessary skills for cooperation with local communities, participation in trans-disciplinary projects with non-engineers for applying the precautionary principles and skills to deal with tensions and sometimes conflicting individual responsibilities.

3.2.2 The challenge of the responsibility for health risks in industrial production environment

The literature survey did not produce a clear answer to the question of whether engineers were only responsible for providing economic benefits to companies and the marketing of new products, or if there were also initiatives on the part of engineers aimed at reducing the harmful impacts of technical processes and products despite potential conflicts with their companies' economic profits. Studies and texts that address the health risks associated with industrial production usually focus on scientific evidence and researchers' reports but rarely comment on the decision-making processes within the firms and the role of engineers in such processes. An example is the book by the medical professor Blanc on "How everyday products make people sick" (2007) which in only a few cases points out the role of engineers.

Regarding bleach[24] production (sodium hypochlorite), Blanc uses the word "industrial chemist": "As one industrial chemist described it 'A product more unfit to go into an inland river would be difficult to conceive'" (Blanc 2007, P. 108).

---

24   Glossary.

In another example using carbon disulfide, he describes emergence of a new engineering discipline, "industrial hygiene", in Italian viscose rayon factories during the Fascist period. The informants, who were required to report on underground organizations in factories, reported on the denunciation of working conditions and carbon disulfide over-exposure and poisoning.[25] The new engineering discipline was not however able to publish its reports in Italy until 1946. If this engineering field were not only a reporting instrument but also a part of decision making, would we read news like this?

> "In September 2000, The American journal of Kidney Diseases reported the case of a man who first came to medical attention at age forty five, suffering from diffuse vascular disease, kidney disease, and neurological complaints. Over the ensuing ten years he worsened, requiring dialysis and developing severe dementia prior to death. He had worked for fifteen years in the spinning department of an unnamed U.S. viscose rayon plant ..."
>
> (Blanc 2007, P. 167)

The industry ignorance and the passive role of society towards the health risks of industrial production have been described more clearly than the role of engineers on that matter. Marx recognized the workers' illness caused by the production environment as the hidden cost of industrial manufacturing. Industrial Societies accepted these costs up to the point where there was enough evidence of great economic loss due to illness and deaths. The health and safety of individuals (in the first line workers) have often been considered as the responsibility of the workers themselves. One example is that of tetrachloroethane ($C_2H_2Cl_4$), used in doping agents as the solvent for cellulose acetate for the production of fighter aircraft between 1910 and 1914 even though British and German researchers had already confirmed its toxicity and the workers' deaths caused by liver failure.

Industry has also repeatedly used benzene, a by-product of coal gasification, for new market opportunities with high health risks for workers although the first benzene incident and the associated workers deaths were documented as early as 1897.

Have engineers neglected the research results on the negative impacts of chemicals and products or processes?

Can engineers today play a more active role in the industry to protect the worker's and consumers?

Looking at such examples we can recognize a very important pattern of technology development and technical progress: The transformation of by-products into new main products through invention without a serious analysis of their possible harmful impacts. Engineers who feel responsible for the health of workers should therefore be aware of new risks not addressed through regulation. Today, they should be aware of the potential risks of new processes and products such

---

25 The first medical notice of carbon disulfide was recorded in 1853 (Blanc 2007, P. 140).

as organic compounds of manganese and lead, which pass into the body more readily than those of inorganic metals in their natural form and are more potently delivered directly to the brain. (see Blanc 2007, P. 253-255)

In conclusion, there is a need for more responsibility on the part of technical universities to make the society aware of industrial health risks. Engineers should play a more active role in protecting the health of workers, consumers and themselves. The chemical industry should require chemical engineers to be aware of the threat of new and old hazards.

A more comprehensive solution would be the integration of hazardous material knowledge in primary and secondary education. Blanc uses the example of carbon disulfide to show how education is responsible for the public ignorance of hazards:

> "How can public education on chemical hazards be so poor that one can still purchase on the Internet a how-to-kit for a classroom exercise called 'the barking dog' experiment, in which phosphorous is dissolved in carbon disulfide that evaporates into the room so that the phosphorous ignites with a 'woof'?"

The education system should not only provide a better understanding of risks and hazards, it also needs to shape the public understanding about different social roles.

3.2.3 Challenges for female engineers

Lessons from history show that there still exists a need for the improvement of technical education for women and cooperation among male and female engineers, as well as the public and private sectors for reforms. The historical description is taken from (Canel et al. 2000).

Business opportunities for female engineers are still a challenging issue. There are still many male-dominated engineering fields. Female engineers are still a minority in the engineering community and in the management positions of engineering firms. This leads to the fact that the idea of female engineers has been excluded from mainstream engineering due to the rarity of women in management positions in public and private institutions. Today there are research projects to analyse the barriers to female engineers' progress and to develop appropriate measures for better career opportunities for women. The recent research project of the Directorate General for Research of the European Union on "Women in science and technology: the business perspective"[26] in 2006 shows that in the European Union, USA, Japan and Turkey, engineering, manufacturing and construction education have the lowest proportion of females in comparison to humanities and arts, social sciences,

---

26 (COM 2006).

business and law, sciences, mathematics and computing, agriculture and veterinary, health and welfare education (with the lowest proportion in Japan (9.2%) and Germany (11.4%)). The report shows that the attractiveness of the job is related among others things to the career opportunities.

Some basic measures to improve the attractiveness of engineering work in companies are coaching, networking and mentoring programs, or measures to improve work-life balance, part-time work, flexible working hours and, to a lesser extent, child care. The cultural and disciplinary differences should also be taken into account.

3.2.4 Problems due to the public perception of the engineering profession

The public perception of the engineering profession and the engineers' self-perception are governed not only by their practice and proficiency but also by their technical education. There is a difference between the engineers' self-perception and the perception by non-engineers.

The definition of the engineering profession is strongly related to the definition of technology. The positive view of technology by engineers leads to a definition of engineering based on finding solutions to technical problems by exercising advanced knowledge and skills. An example of a positive self-perception is the definition by the Royal Academy of Engineering in Great Britain[27]:

> "Engineering is based on the rigorous application of mathematics and science to solve real world problems and design products and processes that benefit society and help to create wealth."

> (RAE 2003)

The positive character of engineering work and its benefits for the society and in particular the engineers' creativity and design capabilities are also claimed in the next citation. The authors remind us of the responsibility of engineers regarding awareness and prevention of negative environmental and social impacts of their designs:

> "Detailed design involves the creation of solutions, product or process designs, or infrastructure designs that meet all the diverse but connected requirements – fitness for purpose, safety, quality, value for money, aesthetics, constructability, ease of use and material efficiency. It does so alongside the minimization of adverse environmental and social impacts, the enhancement of the environment where possible, and the enhancement of quality of life for consumers, workers and neighbours alike. This is a substantial challenge for engineering designers but one that can – with careful thought, creativity, innovation and determination – be delivered for society's benefit."

> (Dodds/Venables 2005, P. 41)

---

27 Http://www.raeng.org.uk/news/publications/list/reports/Future_of_Engineering.pdf, accessed 13 May 2008. The future of engineering research, August 2003.

Dodds and Venables define engineering as a profession with a broad spectrum of capabilities as well as a responsibility for social, environmental and economic benefits. They consider achieving these goals difficult yet possible. In their work they suggest an interdisciplinary view for identifying the potential negative consequences of engineering works:

> "Engineering skills, alongside an ability to work with the many other disciplines involved. It also requires a new view of the world, and a preparedness to adopt new ways of working and thinking about the impacts into the future – negative as well as positive – of engineered products, processes and infrastructure."
>
> (Dodds/Venables 2005, P. 10)

Engineering has been viewed more critically by the non-engineers. There are diverse reasons for this negative perception. The public opinion of engineering is usually influenced by the society's reaction to industrial activities. The engineering profession is sometimes seen to be dull and uncreative. This image that is partly due to the quality of technical education needs to be taken seriously, since it is also a self-perception by some engineers:

> "The issue of the negative image of engineering seems at the base of the problem. Why do we have that image? It clearly starts in college. We work our engineering students through an initial 2 years of dull math and science courses before we let them do the "interesting stuff." Is anybody surprised that we lose 40 percent of those who enrol? I already mentioned that there is an intrinsic tension between our creativity and our conservatism, and that in terms of time measured, we spend more of it on the analytical side than the creative side. I seldom hear engineers talk as I have today about the joy of engineering creativity or associate themselves more with the creative arts than the sciences."
>
> (Wulf 1998)

A negative perception of the profession is also due to the popular belief about engineers' lack of management and communication skills. (see ANSB 2005)

The engineering profession needs its positive public perception to help future engineers to deal with different problems and conflicts during their work, such as the conflicts between the long- and short-term goals or the tension between engineering and science.

3.2.5 Conflicts between the long-term and short-term goals

The time period allotted for the development of technical innovation is getting shorter and shorter due to economic and political short-term strategies. The pressure to generate technical innovations in a shorter time is a reality for engineers today, with known issues as follows:

> "Innovation processes do not depend on the duration of legislative periods or the run time of manager contracts. They can be promoted systematically, but not accelerated for an unlimited period. Innovation remains a spontaneous,

creative process, where even predictable result does not correspond to a certain input as output necessarily."

(Kiper/Schütte 1998)

At the same time innovative enterprises recognize the key economic aspect of long-term goals for innovation.

"It is not easy to find a reasonable and affordable mix for an industrial corporation. Obviously we should have many projects with a horizon of 1-3 years, we should have nearly as many of 3-6 years. We also need a few of 7-10 years, and we have to afford even one or two going beyond 10 years. The latter two are the bridging of the gap to scientific development including universities and to the keeping of a standard of excellence."

(Jucker 1998)

Long-term projects could also be used for a systematic ex-ante impact analysis of technical innovations. It is known that the potential impacts of technical innovation cannot be investigated in a short time scale. Enterprises and policy makers cannot succeed however in pursuing long-term goals to improve the innovations needed without the cooperation of the financial market. The existing financial markets have a destructive effect on the environmental reform of industry due to their economically optimized short-term goals.

An example of a policy for pursuing long-term strategies is addressed in the Environmental Technologies Action Plan (ETAP), a Communication of the European Union in (2004). ETAP introduced performance targets (PTs) as a mean for:

"Setting targets that are both, long-term and visionary as well as perceived as being viable and realistic by many different stakeholders (e.g. consumers, producers and policymakers), is one way to encourage industry to develop and take up environmental technologies. These targets need to be based on best environmental performance, while being realistic from an economic and social efficiency viewpoint, as well as different regional conditions. This means focusing on concrete quantifiable values."

(COM 2004)

Performance targets (PTs) can be defined as quantifiable long-term objectives with measurable parameters representative of the environmental performance of a product, a group of products, a service or a production process.

PTs main characteristics were discussed for the first time in a think-tank meeting in October 2004 in Gothenburg, and were defined as being:

- Representative;
- simple and easy to interpret and communicate;
- scientifically valid, based on data adequately documented and of known quality;
- based on a life-cycle approach;
- capable of being upgraded at regular intervals;

- a clear signal to producers and consumers;
- capable of triggering innovation and/or improving market conditions for environmentally-friendly technologies;
- capable of anticipating and/or accelerating changes and trends over time;
- formulated in a dialogue between many different stakeholders;
- non-legislative in principle, building on the interest of customers and producers to phase out unsustainable technologies.

The multiple aspects of PTs have led to a diffused understanding and different interpretations of this concept. The consensual process of goal setting is one of the main challenges of long-term planning. In addition there are numerous organizational barriers to the changes in the short-term planning system of companies and industry to encompass longer time horizons beyond 5 and even 10 years. The role of engineers should be discussed in depth in this context. Questions for further research are as follows:

- Are engineers who work under short-term plans able to work on long-term projects as well?
- Which skills should be learned during technical education to enable the engineer to be flexible enough to work on both kinds of projects?
- Is it appropriate to have different groups of engineers working on short-term or long-term projects?
- Can engineers recognize the need for different projects in an enterprise with a different time horizon?

Engineers' capabilities and engineering work are usually tightly linked today to scientific work. The discussion of technical education and the necessary skills should therefore include the relationship between engineering and science. The next section briefly addresses this issue.

3.2.6 Links and tension between engineering and science

The history of engineering and technological development during the past centuries shows links as well as tensions between practical and theoretical knowledge. The link was possible mostly for individual engineers such as Hero of antiquity, who also had a deep theoretical knowledge. There was tension between the role of experimental and theoretical knowledge in the understanding of the world and the use of existing knowledge and experience for social development. History shows that different cultures had different understandings and values concerning the practical and theoretical knowledge. These tensions and links were transferred to the relationship between modern engineering and science as well as among engineers and scientists. Engineering sciences today pose a balance of practical and theoretical knowledge from different disciplines. The real practice in different social and political conditions implies however an asymmetrical relationship between different disciplines.

Engineers interact in many different ways with modern science in their work today. Engineering work has in some societies acquired an elite role through a strong technocratic orientation towards basic methods of natural science. This is referred to as Type I relationship between engineering and science. In other societies engineering work is more strongly linked to natural sciences in fields with a root in scientific theories such as gene technology or theoretical physics without an elite role in the society (Type II). Engineering can also be a balanced hybrid of technical knowledge and natural science which is usually presented in chemical engineering and information technology fields (Type III). Engineering can also be considered as a social process. Civil engineering and some fields of information technology are close to this category (Type IV). The fourth type integrates social needs and user perspectives in design and incorporates an intensive communication process for identifying needs, developing solutions and reflecting on the impacts of technical design. This type of engineering process needs an intensive discourse with stakeholders and a high degree of reliance on cooperation among the involved parties. Engineering has in this case a strong relationship to social and human sciences with a broader social context.

In the second and third types of engineering/science relationships there are interdisciplinary bridges that evolve within both engineering and science. Engineers built such bridges between engineering and science during the last decades of the twentieth century. It combined the question of how things happen with the question of how we can use them for a new technical development. Nevertheless a strong tension arises when engineers or scientists pose complex questions. Often scientists and engineers have different interpretations of the same complex problems such as the climate change. As a result of such different understandings, they set different objectives for the same complex questions and use different pathways to achieve their goals.

Engineers search for specific solutions to specific questions or problems. Upon description of the situation and translation of the problem through basic knowledge of mathematics, logic, physics, chemistry, physical chemistry, economics, management, process control, etc., they describe the problem in a new understandable language for engineers and scientists and prepare them for solutions through available theories. These theories are synthesized in engineering handbooks with empirical knowledge on the properties of materials and the functions of systems. Handbooks include predefined pathways for engineers to find abstract solutions. An engineer searches for these abstract solutions with the help of handbooks and specifies them for his or her specific design requirements. The link between engineering and science is present at each step of problem definition, analysis of the situation, experimentation, analysis of results and suggestions for solutions.

But as soon as the question becomes reasonably complex, the engineering hand-books and standards are of no use to engineers. In such circumstances engineers feel the need to gain new knowledge to answer such questions. It takes a period of time for different engineering and science disciplines to make new bridges with each other during the research of new issues. Research and development (R&D) is a key aspect of engineering and science. The interaction between industry and technical universities reflects tensions on setting the priorities for R&D on issues such as climate change. The mainstream industrial engineers have mostly ignored the scientific studies on the urgency of the climate change problem. Political, economic and cultural barriers are not the only hindering factors in developing new and optimal solutions for dealing with the complex problem of climate change. Natural scientists have also not been understood well by industrial engineers. Engineers, who have neglected the impact of conventional production methods on climate change or those who strictly follow the standard feasibility analysis rules of their companies, ignore the interdisciplinary path with the mainstream of natural scientists who view climate change as an urgent problem today.

The feasibility analysis rules are developed in order to evaluate the compatibility of new organizational and technical solutions taking into account the existing structures at companies. Feasibility analysis may act as an inhibitor for radical innovations and prevent the diffusion of new solutions into the production processes, products and services to deal with the climate change problem. This implies that a group of engineers may slow down and inhibit the take-off of innovative organizational and technical solutions with a high potential positive impact on the climate change problem. It is one of the critical issues of engineering profession today that engineers and natural scientists have still different evaluation criteria for setting priorities for their actions.

Another group of engineers who may break down the interdisciplinary bridge with natural science – in the second and third type of relationship – are those who interpret the problem of climate change as a typical problem that can only be solved by new technical solutions. They underestimate the complexity of the natural processes and the requirement for broad political, economic, cultural and social solutions.

The fourth type of relationship between engineering and science could address this broader context. In this type of relationship engineers feel the need to build not only interdisciplinary bridges to natural science but also to social and human sciences, as well as trans-disciplinary bridges to the public, non-scientific and non-engineering communities. Such relationships imply that engineering addresses matters such as precautionary principle, fairness or human rights in their considerations of the complex issue of climate change. In the past such bridges have seldom been built to improve the understandings of the relevant professionals and the public regarding the issues affecting the environment.

This short review of certain relationships between engineering and science shows that technical education in the 21st century has a huge amount of work to do on building the links between engineering, science and society. The relationships between engineering and science are manifold and in part used optimally by engineering. Nevertheless engineering still has many internal tensions that cannot be alleviated in isolation from the society and from its external tensions.

A common internal and external problem for engineering is the obstacles to R&D work on topics with long-term relevance and complexity. Some of these obstacles are described in the next section.

3.2.7 Obstacles to R&D for engineering research

Engineering research follows a specific path within a national research system. The following definition of the UK research community could be considered as a typical European system that exists with different intensities in different European countries.

> "Industry and public research laboratories, together with academic laboratories, used to be the major players in the engineering research community. A small number of private inventors, innovative individuals who perform their own research government policy in areas such as transport, communications, energy and defence exerts a powerful influence on the engineering research community and government needs, therefore, to explore and anticipate the wider implications for UK industry and the supporting research infrastructure when developing such policies."
>
> (RAE 2003)

The research and development system is therefore a heterogenic system with diverse stakeholders with very different interests and influence on setting the objectives and quality of R&D.

Some general organisational obstacles to R&D caused by economic factors are as follows:

- Engineering research in the private sector is fragmented for maximizing the economic profit despite the need to have a review of the product life-cycle in advance.
- The priorities for R&D activities are usually defined based on local requirements. Global problems can be neglected or become lower priorities due to the limited financial resources.
- Private and public funding are more difficult to get for long- and medium term research, although such research is needed for promising technologies in the prototyping stage.
- Ownership of intellectual property rights are often a cause of conflict between industry and technical universities or between large and small companies.

Some specific contemporary obstacles for R&D related to complex problems such as climate change are:

- lack of necessary inter- and trans-disciplinary cooperation and their acceptance by engineers and scientists due to the isolation of the research and scientific systems;
- lack of appropriate scientific methods for combining the new qualitative and quantitative information with the existing knowledge to deal with complex issues such as climate change (neither quantitative, nor qualitative models are able to deal with complex questions such as climate change);
- lack of understanding and application of the precautionary principle to take-off R&D on climate change solutions.

Some of these obstacles are certainly the result of the short history of the study of complex issues such as climate change and the lack of accumulated knowledge on such problems. Some of the obstacles are associated however with the skills learned and the awareness gained by engineers during their technical education.

The obstacles to R&D are also based on the individual views of decision-makers in international companies, or those involved in the local, national and international research systems. A large number of these people are engineers and engineering scientists themselves. Technical education in the 21st century should provide awareness and address these obstacles in the education of engineers.

3.3 Conclusions

This chapter presents a short review of the relevant issues affecting the engineering profession that should be considered in a discussion on the role of the engineering profession in the future. The focus of this book is not to discuss all such topics but to show the diversity of issues and the future needs.

During the Renaissance era the engineering profession demanded a broad spectrum of knowledge together with an artistic character and management skills. As further social and industrial developments occurred, the engineering profession changed by concentrating more on specific technical designs and development tasks. Communication tools and mathematical language became core to the foundation of the engineering profession. The next event was the emergence of industry-focused engineering that shaped the engineering profession into a form more isolated from the rest of the society. Engineers have become industrial staff and have reduced their professional principles to meet only the industrial rules and regulations. The economy has also created a new pole parallel to the political power for shaping the engineering profession.

Investors' interest in industrial mass production and hierarchical production systems is perhaps one of the main factors causing the isolation of engineering work hampering their communication with users and preventing reductionism in technical design for efficiency. Lessons from history show that engineers had to use reductionism and concentrate on specific problems to generate ideas for technical solutions. The American history of independent engineers is full of examples such as Edison who established inventory factories and a large number of inventions of products that were later manufactured using mass production systems designed by Henry Ford (American engineer and automobile manufacturer, 1863-1947). The risks associated with products and production systems remained unaddressed for a long time in such systems. Risks and environmental impacts of isolated industrial designs could not be easily addressed, analysed and remedied by the same system because of the specialisation required for the process.

Inter- and trans-disciplinary works are needed today in engineering design to deal with complex issues such as climate change. Technical solutions constitute in these cases often only one element among many other non-technical solutions. This is however against the technocratic thinking that believes in technical solutions even to technology-induced problems. Technocrats emphasize technical efficiency as the core answer to all needs and deficiencies. In addition, engineers and scientists also need to deal with tensions within the engineering profession and between engineering and science. Both sides have to build links to one another to be able to work effectively on complex issues such as climate change.

History does not show many independent engineers who were socially committed and skilful enough to provide society with an insight into the potential problems of new technical solutions. The engineering profession and technical education have neglected the importance of the ex-ante critical view of technical solutions in the last two centuries.

The missing link was perhaps an independent advisory engineering association or an international non-governmental organisation for engineering which would have kept the industry focused not only on the predictable negative effects but also on the unintended potential long-term impacts of new or existing technologies. A group of social scientists and philosophers such as Habermas and Feenberg have emphasised the importance of communication and information exchange about technical possibilities in the last decades:

The German philosopher Jürgen Habermas states that

> "Needs must be interpreted in the light of values and cultural meanings before they can guide action. Habermas: On the one hand, the horizon of values in a society guides scientific research, on the other, value convictions persist only insofar as they are connected to potential satisfaction through instrumental action. Consequently technology cannot be value-neutral. Practically relevant scientific achievements must be subjected to free public discussion to make possible 'a dialectic of enlightened will and self-conscious potential' that both allows new technologies to alter public self-understanding and lets that self-understanding determines the course of future research. Insofar as such discus-

sion is governed by the 'unforced force of the better argument,' it yields decisions on ends that are rational in a sense decisionists failed to recognize."[28]

(Habermas 1970, P. 73 – see also McCarthy 1989)

A rational engineering profession takes part in a dialogue with society and discusses society's needs as well as feasibilities and acceptability of technological solutions. Within such an engineering profession, free from the dogma of technical efficiency, individuals can enjoy flexibility and freedom to improve the level of technical standards for environmental protection and social considerations (see Veak 2006).

The new needs for change in the engineering profession identified through this literature survey are classified into three main categories as follows:

- *organisational changes* for inter- and trans-disciplinary work on complex problems;
- *conceptual changes* for the integration of an ex-ante impact assessment of technical design in engineering work;
- *behavioural changes* to emphasise the individual capabilities of engineers to communicate with the public and an improvement of the self-understanding by individual engineers of their global responsibilities beyond their company and industry rules.

In the next chapter the author focuses on individual engineers' opinions about the changes necessary in their profession. The chapter shows the results of a survey conducted in 2006. For this work it was necessary to compare the answers given by individual engineers from different cultures with the results of the literature survey. The goal in particular was to know whether the interviewees saw any need for changes in technical education in comparison with their own education and whether they had any additional factors in mind affecting success as related to needs for change.

---

28 Habermas: http://www.bookrags.com/research/habermas-jrgen-este-0001_0002_0, accessed 13 May 2008.

4. Needs for the future engineering skills based on engineers' opinions

Marcus Virtuvius Pollio wrote about the need for a balance between knowledge and skills in the work of architects in the first century B.C. The architect was the usual nomenclature for engineers in that epoch:

> "... architects who without culture aim at manual skill cannot gain a prestige corresponding to their labours, while those who trust to theory and literature obviously follow a shadow and not reality. But those who have amatered both, like men equipped in full armor acquire influence and attain their purpose."

> (Rae/Volti 1999, P. 22)

He specified this knowledge and skills further and called for a broad understanding of architects' skills:

> "... He should be a man of letters, skilful draftsman, a mathematician, familiar with historical studies, a diligent student of philosophy, acquainted with music, not ignorant of medicine, learned in the responses of jurisconsults, familiar with astronomy and astronomical calculations."

> (Rae/Volti 1999, P. 22)

How do we define engineers' skills today? And what do we expect from the engineering profession in the future?

Today we arrive again at the need for a broad understanding of engineering: "It has been asserted many times that much of the operational development of sustainable systems throughout this planet will fall to broad based, systemically educated engineers" (Nichol 2006).

There are however different opinions about the skills required by engineers'. Studies by various engineering associations, conferences and declarations use different terms when addressing these needs.

A recent study by the American National Academy of Engineering, called "The Engineer of 2020, Visions of Engineering in the New Century" (NAE 2005), presents a list of "opportunities and challenges for the 21st century". The study identified the following challenges and trends for the engineering profession.

- a global population approaching 10 billion with a steadily aging demographic and a growing demand for diversity in the engineering workforce;
- an imperative for "sustainability" in the face of global population growth, industrialization, urbanization, and environmental degradation;
- an increased focus on managed risk and assessment with a view on public privacy, safety and security;
- the globalization of economic systems and the interconnectedness of its component parts;

- the accelerating pace of technological advances, including the increasing importance of information technology, communications science, and biological materials and processes in engineering;
- growing concerns about the social and political implications of rapid technological advances and their uneven application among different constituent groups (e.g., the digital divide, medical ethics, etc.);
- the diminishing half-life of engineering knowledge in many fields;
- the growing complexity, uncertainty, and interdisciplinary foundations of engineered systems;
- the growth of the "services-based" component of the economy;
- the increasing number of engineers working in non-traditional areas that require technological competence and/or fluency (e.g., management, finance, marketing, public policy, etc.);
- the shift in engineering employment from large companies to small and medium-sized companies and the growing emphasis on entrepreneurialism;
- increasing opportunities for incorporating technology into the education and work life of engineers."[29]

The list above shows on one hand challenges such as diversity in engineering, industrialisation, urbanisation, globalisation, environmental degradation, acceleration of technological development, and the growing complexity of engineering systems; and on the other hand it clearly reveals the need for risk assessment, reflection of impacts, more non-engineering skills and flexibility of the engineering profession. Moreover it identifies trends in engineering working more closely with users.

The British Royal Academy of Engineering combines improving the environmental performance with the economic dimension of engineers' obligations as follows:

> "Engineers should ensure products, services and infrastructure for a more-sustainable life with improved environmental performance, while these products, services and infrastructure are competitive in their marketplace and, ideally, the most competitive."
>
> (Dodds/Venables 2005, P. 10)[30]

In all these new definitions for engineering skills we see that in comparison to the past centuries engineers today are considered responsible for more than economic growth.

---

29 Prepared by the American National Academy of Engineering Committee on Engineering Education, 2005.

30 Other principles of the guideline are: "drive down the adverse environmental and social aspects of engineered products, services and infrastructure; dramatically improve their environmental performance; improve the contribution of engineering products; services and infrastructure to a high quality of life, help society to move towards a significantly more-sustainable lifestyle" (Dodds/Venables 2005, P. 10).

"State engineers in the late seventeenth century in France were to regulate and direct the country's entire technical infrastructure, in order to further economic growth and ensure better living standards."

(Cross 1998, P. 129).

Today, the engineering profession is faced with consequences of techno-centric decisions and economic rationalism by industrial societies, which incorporate risks without accountability. This problem is also discussed in the theory of "risk society" by Ulrich Beck. According to Beck's concept of "risk society" organizations within industrial societies are no longer capable of dealing with the global and unpredictable risks they impose upon their societies. The new questions on the relationship between nature and society and the presence or absence of borders within a society are those that impact on the understanding of relationships among society, science and technology. These questions bring about new discussions on engineers' required responsibilities and should be addressed in the change process of technical education in the 21st century. In the next sections of this chapter we will discuss the significance of the engineering profession's new roles and responsibilities.

4.1 Diverse Engineering roles according to different interpretations of technology with sustainability

Cultural versus Ecological sustainability:

Debates on technological advancement in sustainable development exhibit both technical and socio-economic dimensions. They lead to very different understandings of engineering's role. Numerous interpretations of sustainable development exist that consider different objects to be sustained through various means to achieve sustainability.

In his book Davison (2001) presents the dissent on different interpretations of sustainable development and contests meanings of sustainability related to technology development. Considering the diverse moral and political backgrounds of discussions on sustainability Davison differentiates between "cultural sustainability" and "ecological modernization" (2001). Ecological modernisation is linked to the post-war environmental crises and to the belief that technical solutions can solve environmental problems through eco-efficiency. His "Discourse on cultural sustainability" considers the cultural and moral policies and values as the basis for development and suggests a type of technology development that does not view the technology as the main ingredient for progress. Davison claims that the limits imposed upon cultural sustainability are due to the absence of direction for social development and clear definitions of objectives for development. This has probably caused the consideration of economic growth as a core objective for ecological modernisation towards sustainable development and giving economic advancement a dominant role in comparison to the social and cultural aspects. "Ecological modernisation" supports the idea of eco-efficiency and clean technologies as means for growth and sustainability. In the eco-efficiency concept technical innovations are means for resources management and energy efficiency, as well as for recycling, reduction of waste and prevention of toxicity and other environmental hazards.

Ecological modernisation mainly addresses the short-term strategies for "win-win" effects. "Win-win" effects are a vague definition of situations where economic goals can be achieved combined with ecological or social benefits. The problem with these definitions is that the levels and nature of the benefits being evaluated and the corresponding time frames are usually unclear. The reports on sustainable development such as "Our common future" known as the Brundtland report (1987) claim the engineers' responsibilities for developing environmentally sound and clean technologies which are an economic drive for sustainable development. The eco-efficiency or eco-modernisation concepts focus on new and efficient production processes and products. According to this point of view there should be a change in norms and standards to allow acceptance of new efficient technical products, processes and services. This option does not emphasize the individual role of engineers and is a technocratic view that views the solution of problems only through development and implementation of new technologies. This type of engineering responsibility therefore has a minimal impact on changing the behaviour of individual engineers.

Another point of view considers technology as a means of sustainability in a frame-work of social and environmental preconditions. This perspective points towards a change in boundary conditions for the engineering profession in a way that would also change the engineers' behaviour. An example of this concept is given in the Fourteenth Session of the UN Commission on Sustainable Development (CSD-14), 01 May 2006 – 12 May 2006 in New York. The Science & Technology NGO Caucus Statement (Swati Save) states:

> "The application of Science and Technology for achieving sustainable energy resources and services for the rural and peri-urban community should take into consideration the following:
> 1) women's participation – particularly emphasis on basis primary education;
> 2) promoting cleaner energy;
> 3) using local resources;
> 4) appropriateness to resources and skills level;
> 5) government, and private sector initial funding and sponsorship;
> 6) an integrated energy framework and cultural bias."[31]

---

31 Http://sustsci.aaas.org/files/CSD-14%20Sci%20&%20Tech%20final%20statement.doc, accessed 16 June 2008.

These preconditions demand non-technical and non-economic considerations in the application of technology. They imply a higher degree of consciousness about the risks and uncertainties associated with the impacts of engineering tasks. Examples of such engineering works can be found in civil engineering where engineers have direct contacts with their local communities. For the majority of engineers their activities are limited to their obligation towards quantified standards and norms within their firms.

A number of examples of non-economic values for engineering tasks of individual engineers are given in the international conference on Engineering Education in a Sustainable Development (EESD 2006) in Lyon.

The need for professionals to look behind and beyond the borders of their companies and think critically about the impact of technical systems is expanding:

> "What is needed, are professionals who can look beyond current developments, outside prevailing systems, to unchain new developments toward sustainability."

> (Werk et al. 2006)

The EESD 2006 in Lyon included discussions within a scientific community with emphasis on the technical dimension of sustainable development. The author compared the issues regarding the engineering profession challenges, and future trends in this conference against the results of the survey which will be presented in this chapter. For a more comprehensive representation of other engineering communities' opinions the author also used the proceedings of SEFI 2001 Conference in Copenhagen that focuses on the general changing paradigm of engineering education[32]. Topics discussed in these proceedings included the need for new engineering competencies for critical thinking, social and environmental responsibility, challenges introduced by new business dynamics and application of information and communication technologies in engineering activities.

The declarations by engineering communities are another source of demand for changes. For the purpose of designing the survey questions the author used the Declaration of Barcelona (2004)[33] which stands for a broader view of engineering professions and technical education. The Declaration of Barcelona (2004) is one of the most explicit forms of self-obligation for engineering, set by an engineering community with a focus on "sustainable development" as a social value for engineering.

---

32 EESD 2006: (Subai et al. 2006), SEFI 2001: (Graaff et al. 2001), The European Society for Engineering Education, SEFI, is the largest network of higher engineering institutions and educators in Europe and contributes to the development and improvement of the position of engineering education in society. (for paradigm see glossary).

33 Http://www.eesd2006.net/index.php?option=com_content&task=view&id=42&Itemid=96, accessed 16 June 2008.

4.2 The Declaration of Barcelona (EESD 2004)

The Declaration of Barcelona published in the EESD 2004 proceedings states that today's (and future) engineers must be able to:

"• understand how their work interacts with society and the environment, locally and globally, in order to identify potential challenges, risks and impacts;
• understand the contribution of their work in different cultural, social and political contexts and take those differences into account;
• work in multidisciplinary teams, in order to adapt current technology to the demands imposed by sustainable lifestyles, resource efficiency, pollution prevention and waste management;
• apply a holistic and systemic approach to solving problems and the ability to move beyond the tradition of breaking reality down into disconnected parts;
• participate actively in the discussion and definition of economic, social and technological policies, to help redirect society towards more sustainable development;
• apply professional knowledge according to deontological principles and universal values and ethics;
• Listen closely to the demands of citizens and other stakeholders".[34]

The Declaration of Barcelona calls for more responsibility on behalf of individual engineers through comprehensive system-oriented thinking and impact assessment to reduce negative effects of technologies. The declaration, furthermore, demands an active role of engineers in the movement towards sustainable development.

4.3 Method of the explorative survey

Through comparative research a broad list of challenges has been identified for engineering in the future. A plan for changes in technical education solely based on these results could however lead to misinterpretation of priorities due to the diversity of information sources and references. In order to identify key engineering skills the results of the comparative research have been contrasted against the results of an explorative survey which includes questions regarding past, present and future engineering key skills. This survey addressed engineers from various countries with very different backgrounds. This section presents the methodology used in this survey.

The explorative survey was performed during the period of June through to November 2006 and gathered ideas from engineers with a wide spectrum of knowledge and engineering experience (electronics, building, surveying, mapping, construction, chemistry, computer, etc.) and from natural scientists with previous experiences at technical universities.

---

34 Http://www.eesd2006.net/index.php?option=com_content&task=view&id=42&Itemid=96, accessed 16 June 2008.

The survey questionnaire (see Appendix B) contained elements that universities usually use for their internal surveys[35] with additional questions relevant to the topic of engineering within sustainable development (such as engagement of students in international projects on environmental and social issues and local initiatives during their education). Other topics discussed in mainstream engineering communities such as common engineering education that was addressed in a project by the American National Academy of Engineering (see Clough 2005; NAE 2005) were also covered. In some cases it was possible to describe and pose the questions to the interviewees face to face or over the phone. In other cases the questionnaires with short descriptions and the background of the study were sent by e-mail to the respondents. It was often necessary to provide information regarding the concept of sustainable development to the respondents of the survey.

The group of respondents included 44 individuals who were mostly engineers with diverse cultural and engineering backgrounds and different levels of experience in research, teaching, management and business. The respondent group included both international experts with a wide spectrum of activities as well as engineers engaged in pure engineering tasks. Some respondents had worked on development of concepts for engineering education in sustainable development. Respondents included women and men of different ages. 14 women and 30 men from different countries and backgrounds took part in the survey (more than 300 people were asked to take part in the survey).

The survey results provide a valuable qualitative source of data for analysis in this book and will not be analyzed quantitatively. The results are used to add insights to the findings from the comparative research. The explorative survey provides an up-to-date pool of information in the language of engineers. Respondents described their past and present experiences and their ideas about future requirements for engineering work and technical education.

The survey results are provided anonymously in this analysis. A list of questionnaires with corresponding numbers is given below in Table 1:

Table 1: Information on interviewees; the background is based on information from the Q1-Q3 of the questionnaire

Nr.	Country	Gender	Background
1	Austria	W	Chemical engineering, Biochemistry, Biotechnology, 8 years of experience
2	Austria	W	Chemical engineering, young engineer
3	Austria	W	Chemical engineer, 7 years of experience
4	Finland	W	Chemical engineer, EESD 2006 participant
5	Poland	W	Chemical engineer, EESD 2006 participant

---

[35] Example: (CEAS-UCB).

Nr.	Country	Gender	Background
6	Austria	M	Chemical engineering at different universities, 20 years of experience
7	Austria	M	Chemical engineer, 10 years of experience
8	Austria	M	Chemical engineering, 20 years of research and education experience, EESD 2006 Participant
9	Austria	M	Chemical engineer, about 5 years of experience
10	Austria	M	Chemical engineer, 15 years of experience
11	Austria	M	Chemical engineer and manager, 30 years of research and education experiences
12	Netherlands	M	Chemical engineer, background also in economics, 25 years of experience, EESD 2006 participant
13	Iran	M	Chemical/Environmental Engineer
14	Sweden	M	Civil engineer, 30 years of experience
15	USA	M	Civil engineer with background in Electronics, Study in different countries, 20 years of experience
16	USA	M	Computer engineer with industrial engineering background
17	USA	M	Computer engineer, young engineer
18	France	W	Engineer: Cleaner Technologies (CT), Waste Science and Techniques, 8 years of experience, EESD 2006 participant
19	USA	M	Electronic engineer, Computer science, 20 years of experience
20	USA	W	Electronic engineer
21	USA	M	Electronic engineer and instructor, 30 years of experience
22	USA	M	Electronic engineer, 15 years of experience
23	USA	M	Electronic engineer, 20 years of experience
24	USA	M	Electronic engineer and MBA, about 10 years of experience
25	USA	M	Electronic engineer
26	USA	M	Electronic engineer
27	USA	M	Electronic engineer, 20 years of experience
28	USA	M	Electronic engineer, study in different countries, young engineer
29	Canada	W	Electronic engineer, study in different countries, young engineer
30	Austria	W	Electronic engineer, 20 years of experience
31	USA	M	Industrial engineer, computer engineer, study in different countries, 15 years of experience
32	USA	M	Industrial engineer, applied statistics, 25 years of experience
33	USA	M	Materials engineer, study in different countries, 12 years of experience
34	Austria	W	Mechanical and chemical engineer, 10 years of experience
35	Austria	M	Mechanical engineer, about 10 years of experience, EESD 2006 participant
36	USA	M	Mechanical engineer, study in different countries, 23 years of experience
37	USA	M	Mechanical engineer, study in different countries, 25 years of experience

Nr.	Country	Gender	Background
38	Netherlands	W	Natural scientist, Biochemistry, many years experiences, EESD 2006 participant
39	Spain	W	Physicist, teacher in a school of architecture, 25 years of experience, EESD 2006 participant
40	USA	M	Ph.D. Physics, study in different countries, 20 years of experience
41	Austria	W	Planning Engineer, young researcher, EESD 2006 participant
42	Sweden	M	Process Engineering could be also a kind of Chemical engineering
43	Netherlands	W	Science and policy, EESD 2006 participant
44	USA	M	Surveying and Mapping, engineer, 22 years of experience

The questionnaires are numbered 1-44 to help readers in referencing detailed information across different tables more easily. The summary of results and conclusions is presented in the last section of this chapter (section 4.8).

The complete questionnaire is shown in Appendix B and it starts with questions regarding the background of respondents (see Appendix B, Q1, Q2, Q3).

Questions 4 to 9 are taken from the Bologna process[36] and Declaration of Barcelona which are regarded as two important baselines for technical education in the 21st century.

Respondents were asked about their experiences and how important they considered the factors listed.

An example for Q4 is given below:

4a	Have you had off-campus (industry) learning processes during your education at the university?	yes	no			
		☐	☐			
4b	How strongly should the external learning process be emphasized at technical universities?	very strongly	strongly	weakly	very weakly	not at all
		☐	☐	☐	☐	☐

Questions 5 to question 9 are like question 4 based on:
• interdisciplinary courses and projects during education (Q5);
• participation in project initiatives on local issues during education (Q6);
• participation in international projects on environmental or social issues during education (Q7);
• involvement in research activities during education (Q8);
• involvement in teaching activities during education (Q9).

---

36 Glossary.

The next part of the survey included two questions on relevant success factors for respondents themselves and their ideas about success factors for future engineers:

Q 10: Please name three skills you have learned at the university which have been important for your work-life success.

Q 11: Please name three skills which will be important for the work-life in the future and should be learned at technical universities.

It was also possible to name more than three factors in answering these questions. Some interviewees used this opportunity to expand on their points of views.

The last question (Q12) gave the respondents the chance to explain their suggestions on technical education in 2020 within the context of sustainable development.

The survey design aimed at enabling different people to present their ideas in different ways. The result of the survey and the analyses of information are presented in the next sections. The results of question 10, 11 and 12 are fully documented in Appendix B.

4.4 Factors important to engineers (Q4 to Q9)

Question 4 is related to the off-campus experiences during education. Table 2 shows that not all respondents had experiences of this kind.

Table 2: Results of Q4; 43 respondents answered this question

Nr.	Question	Own experiences of interviewees	
		yes	no
Q4	Have you had off-campus (industry) learning processes during your education at the university?	26	17
How strongly should the external learning process be emphasized at technical universities?				
Very strongly	Strongly	Weakly	Very weakly	not at all
14	25	3	1	

The results shown in Table 2 indicate that the majority of the group, with or without their own off-campus experiences, consider this factor very strongly or strongly important. This question is also an indicator for the need for more practical experiences and the need for problem-based learning balanced with theoretical knowledge. This observation is consistent with the results of the comparative research.

Another such key factor identified through desk research is addressed in the next question. Question 5 asks respondents about their experiences regarding inter-disciplinary courses and projects and the importance of inter-disciplinary experiences for future engineers.

Table 3: Results of Q5; 43 respondents answered the first part of the question; 42 respondents answered the second part of the question

Nr.	Question	Own experiences of interviewees	
		yes	no
Q5	Have you had interdisciplinary courses and projects during your education?	23	20
How strongly should the interdisciplinary courses and projects be emphasized at technical universities?			

Very strongly	Strongly	Weakly	Very weakly	not at all
17	21	4		

Table 3 indicates that this group considered interdisciplinary courses and projects as very strongly and strongly important. Even those without personal experiences with inter-disciplinary projects during their education viewed such projects as strongly important for future technical education. This observation is in line with the results of the desk research, analogous to the previous factor "off-campus experiences".

The next question on experiments with local issues during education paints a different picture. Most respondents did not have their own experiences with project initiatives on local issues.

Table 4: Results of Q6; 43 respondents answered the first part of the question; 42 respondents answered the second part of the question

Nr.	Question	Own experiences of interviewees	
		yes	no
Q6	Have you participated in project initiatives on local issues during your education?	10	33
How strongly should projects on local problems be emphasized at technical universities?			

Very strongly	Strongly	Weakly	Very weakly	not at all
8	21	9	3	1

Table 4 shows that only a small group of respondents considered these experiences very strongly important. Half of all respondents categorized these experiences as strongly important. There were also sceptical opinions on the importance of local initiatives for technical education. It is important to note that mainstream engineering follows the trend for efficient engineering education to prepare engineers for international careers. Local issues are often not relevant in such programs. Other reasons may exist explaining the scepticism about engaging students in local project initiatives – for example students should have the opportunity to select their own project topics regardless of whether or not the topics include local issues. The respondents' answers showed no trend based on nationality differences.

The replies to the next question pointed out a central factor for sustainable development. Question 7 asked about international experiences and language skills on environmental and social issues during education that are important factors in the European Bologna process.

Table 5: Results of Q7. 43 respondents answered this question

Nr.	Question	Own experiences of interviewees	
		yes	no
Q7	Have you participated in international projects on environmental or social issues during your education?	10	33
How strongly should projects on local problems be emphasized at technical universities?			
Very strongly	Strongly	Weakly	Very weakly	not at all
6	23	14		

Most of the European and some U.S. respondents considered participation in international projects on environmental and social experiences as very important for future engineers' success. A third of respondents considered international projects during education as weakly important. The scepticism regarding the level of importance of this factor for the future may seem strange at first. It is important to notice that the majority of respondents who had participated in international projects viewed this factor as strongly important and important. This sub-group perhaps had a clearer interpretation of such international experiences. The other part of the group might have interpreted the international projects on environmental and social issues very differently, and therefore not as relevant. Despite the lower evaluation of this factor the result of this question might match well with the results of the desk research on the vitality of sustainability and the need for consideration of environmental and social impacts as a challenge for future engineers.

In the next two questions respondents were asked about the involvement of students in two key activities at technical universities. Question 8 addressed the students' involvements in research activities during their education. This is one of the key issues that could indicate the importance of research with students' participation at technical universities.

Table 6: Results of Q8; 43 respondents answered this question

Nr.	Question	Own experiences of interviewees	
		yes	no
Q8	Have you been involved in research activities during your education at technical universities?	33	10
How strongly should projects on local problems be emphasized at technical universities?			
Very strongly	Strongly	Weakly	Very weakly	not at all
22	16	5		

Table 6 shows that many respondents had their own experiences with research activities during education at technical universities and the majority considered it as very strongly or strongly important in the future. The response of this sub-group emphasises the need for the multi-functionality of technical universities. Almost in the same manner, however, contrary views exist which demand only educational functions for technical universities.

The next table shows the results of the survey for question 9 on teaching activities during education. The results for questions 8 and 9 could be seen as complementary pieces of information on different functions of technical universities. At the same time these two questions address the learning by doing at technical universities for those engineers who are going to pursue a scientific or research career.

Table 7: Results of Q9. 40 respondents answered this question

Nr.	Question	Own experiences of interviewees	
		yes	no
Q9	Have you been involved in teaching activities during your education at technical universities?	20	20
How strongly should students be involved in teaching at technical universities? (Tutors, …)			
Very strongly	Strongly	Weakly	Very weakly	not at all
7	16	15	2	

Table 7 shows that half of the respondents had their own experiences with teaching activities during their education but a sub-group of European respondents considered it as only weakly and very weakly important for the future. Only one American respondent considered this factor as weakly important. According to the explorative survey, the majority of respondents perceive teaching experiences to be less important in future than research during education. The desk research showed the same sceptical trend for direct emphasis on teaching experiences and tutorial practices during technical education.

Some interviewees from the USA said in direct talks that teaching experiences could have a strong impact on engineers' general skills. According to newly identified trends for involvement of engineers in different jobs and their work in small and medium-sized engineering companies in the USA[37], it can be concluded that teaching experiences might provide a deeper understanding about communication of technical knowledge. Therefore temporary teaching experiences can not only be considered as training for those who plan to follow an academic career but also as a useful training measure for all engineering students.

The analysis of responses to question 9 shows that a group of European respondents, who viewed teaching experiences as less important, are participants of the

---

37 See ANSB 2005.

international conference EESD 2006 that had its focus on engineering education in sustainable development.

The previous two questions (Q8 and Q9) imply together a definition of different functions of technical universities and the role of students to learn and be engaged in research activities.

The next section provides an overview of the results of questions 4 to 9. These findings led to recognition of some key success factors for future engineers.

4.4.1 Summary of the results of Q4 to Q9

The respondents have evaluated the following 5 of 6 success factors as having a fairly similar level of importance as present and/or future success factors. These factors are:

Experiences with:
- off-campus projects;
- interdisciplinary courses and projects;
- projects at local level;
- projects at international level on environmental and social issues;
- research activities at technical universities.

Teaching experience during technical education was seen as a success factor by the majority of American and some European respondents. The responses mostly differed from the respondents' own past experiences.

Teaching experiences were considered by some European respondents as a weakly important success factor. The list of respondents for this question shows that the majority who had this sceptical opinion about the importance of teaching experiences were EESD 2006 participants. With respect to the other topics covered in the survey these respondents did not form a specific sub-group. The author interprets these similar points of views as an indicator of context-related responses of participants of EESD 2006 to special success factors related to sustainable development. This result could be interpreted as less emphasis on teaching activities as a specific factor for engineering work within sustainable development.

Another conclusion from the responses of this sub-group to Q9 and their sceptical opinion regarding the strong importance of teaching experiences could imply a focus of these respondents on the practical dimension of engineering work in sustainable development. Nevertheless, teaching could also be interpreted as an underlying general success factor for engineers in every context.

The results of questions 4 to 9 should be analysed in combination with the next questions where respondents had the opportunity to define skills that they regard as their own success factors. Questions 10 and 11 addressed the differences between the past and future. Question 12 represented suggestions on the future in the context of sustainable development.

4.5 Key skills for present and future engineering work-life (question 10 and 11)

Question 10: Please name three skills you have learned at university which have been important for your work-life-success.

Question 11: Please name three skills which will be important for the work life in the future and should be learned at technical universities.

The responses to questions 10 and 11 are given in Appendix B. They represent answers to "success factors for own work-life" of respondents (question 10) and "success factors for future engineers" (question 11).

These questions serve to identify success factors in addition to the desk research results and whether different sets of key skills exist for the present and future.

A first look at the results showed that each respondent listed a different set of factors in answering questions 10 and 11.

An entirely similar combination of factors could not consistently be found in the responses to the questionnaires. Some statements such as "communication skills" or "management skills" were, however, repeated in different questionnaires.

There were 125 statements on the "success factors for own work-life" (question 10) and 132 stated factors on the "success factors for future engineers" (question 11) that were divided into 11 different categories (types). The detailed discussions on pertinent success factors in each category are given in the next section.

Section 4.5.1 starts with a presentation of responses to question 11 on future success factors. These results will be compared to responses on "success factors for own work life" in section 4.3.2.

The list of categories of future success factors in responses to question 11 ordered with respect to their priorities is as follows:
- "Communication and presentation skills";
- "Management skills";
- "Basic engineering skills";
- "Cooperation skills";
- "IT skills";
- "Cross-disciplinary work skills";
- "Reflexive knowledge";
- "International experiences and language skills";
- "Practical knowledge gained from learning by doing";

- "Capability of thinking and acting beyond the borders of the company on global and local level";
- "Capability of critical thinking and taking responsibility".

A group of personal factors are classified to "Miscellaneous factors".

The category "Miscellaneous factors" with some personal skills is not at the same level as the categories "Cooperation skills" or "Capability of critical thinking and taking responsibility". This asymmetrical terminology was necessary in order to integrate as many terms as possible from the respondents' statements. The interrelationships of these categories will be discussed later in the conclusion of this chapter.

Table 8 shows that another priority list can be defined for the success factors listed in question 10 with respect to the respondents' own work life. It shows emphasis by the respondents on their own success factors and on future key skills. The contrast between the responses to question 10 and 11 can be seen as an indicator for needs of new engineering skills.

Examples for "Miscellaneous factors" for future success factors are:
- Dealing with multiple issues at the same time;
- Open and broad minds;
- Analytical thinking;
- Problem-solving skills;
- How to interpret knowledge to practice;
- Creativity;
- Self confidence;
- Career development;
- Basic environmental knowledge.

"Miscellaneous factors" for their own success are listed as:
- Analytical and logical thinking;
- Problem-solving skills;
- Persistence/endurance;
- Independency;
- Flexibility;
- Self-confidence;
- Focus and concentration;
- Life-long learning;
- Learning skills;
- Feedback;
- Patience.

Table 8: Comparison of priorities of identified factor groups for the "own success factors" of respondents and "future success factors"

Factor Group	Future success factors named by respondents (Q11)	Own success factor of respondents (Q10)
Communication and presentation skills	[2], [4], [6], [8], [10], [12], [13], [14], [16], [17], [19], [21], [27], [28], [29], [32], [33], [34], [38], [41], [42], [44]	[1], [6], [8], [9], [11], [15], [21], [25], [29], [33], [38], [41]
Management skills	[2], [4], [5], [6], [7], [13], [16], [17], [21], [27], [28], [31], [32], [33], [41], [42], [44]	[2], [4], [5], [14], [16], [35], [36], [42]
Basic engineering skills	[5], [6], [10], [11], [12], [21], [28], [31], [32], [34], [38], [40]	[5], [6], [10], [11], [12], [14], [17], [19], [25], [28], [31], [32], [33], [34], [35], [37], [38], [40], [41]
Cooperation skills	[1], [2], [3], [4], [9], [10], [14], [16], [18], [26], [30]	[3], [4], [13], [14], [18], [21], [22]
IT skills	[19], [31], [34], [36], [37], [38], [40], [41], [44]	[17], [19], [34], [37], [40]
Cross-disciplinary work skills	[6], [11], [12], [15], [25], [26], [27], [29], [35], [38], [39]	[4], [34], [36], [39], [41]
Reflexive knowledge	[8], [12], [15], [18], [19], [25], [36], [37]	[8], [12], [38], [39]
International experiences and language skills	[3], [4], [7], [8], [14], [34], [39]	[9], [40], [42]
Practical knowledge gained from learning by doing	[18], [30], [32], [33], [35], [41], [44]	[5], [13], [16], [31], [32], [33], [37]
Capability of thinking and acting beyond the borders of the company on a global and local level	[4], [9], [14], [15], [22]	[4], [27]
Capability of critical thinking and responsibility	[13], [36]	[26]
Miscellaneous factors	[1], [9], [11],[22], [29], [42]	[1], [2], [4], [6], [7], [8], [9], [11], [15], [17], [18], [21], [22], [23], [25], [26], [27], [28], [29], [30], [31], [33], [35], [36], [39], [41]

In the following section the responses to question 11 on future success factors and key skills are analysed.

4.5.1 Specification of engineers' future key skills

In this section the responses to question 11 on "future success factors" of engineering work are discussed for each of the 12 factors identified by the survey. The answers to each category will be compared to factors identified in the conference proceedings SEFI 2001 and EESD 2006[38]. The main purpose of this comparison is to find similarities and differences among the identified success factors and the factors discussed within various engineering communities. The EESD 2006 in Lyon presents the discussions within a scientific community that emphasize the technical dimension of sustainable development. The proceedings of SEFI 2001 Conference in Copenhagen focuses on the general changing paradigm of engineering education.

For analysing the responses from different contexts of engineering challenges perspectives, each category is followed by a short comment which will be discussed further at the end of this section.

The category "Communication and presentation skills" includes the following factors:

- Social networking skills;
- Social and soft skills for handling people's issue and effective communication;
- (public) Presentation skills;
- Rhetorical skills;
- Effective and impact-full technical writing.

The "Communication and presentation skills" factor group has a high priority in the list (Table 8: mentioned by 22 of 44 respondents).

Similar factors in EESD 2006 are social and soft skills for handling people's issues, effective communication and public presentation skills. The SEFI 2001 conference proceedings contain a group of factors on effective communication, public presentation skills, rhetorical skills, effective technical writing. "Communication and presentation skills" can therefore be regarded as important key skills for future engineering work.

---

38 SEFI 2001: (Graaff et al. 2001), EESD 2006: (Subai et al. 2006), The European Society for Engineering Education, SEFI, is the largest network of higher engineering institutions and educators in Europe and contributes to the development and improvement of the position of engineering education in society.

The category "Management skills" includes the following factors:
- Management skills;
- Project management;
- Time management;
- Organization skills;
- Resources management and planning including out-sourcing;
- Knowledge of marketing;
- Leadership;
- Basic understanding of business administration;
- Knowledge of intellectual property;
- Executing projects.

Table 8 shows that 17 out of 44 respondents with very different backgrounds expect a strong future need for different management methods and understanding of business administration in the engineers' work. The majority of this sub-group has also stated communication and presentation skills will be important. Management skills are strongly connected to communication skills for respondents within this group.

Similar factors in EESD 2006 include effective leadership as well as social and managerial knowledge. The SEFI 2001 conference proceedings include contributions that emphasise management and business skills. Both the desk research and the survey showed that management skills are considered as general and fundamental future key skills for engineers.

The category "Basic engineering skills" includes the following factors:
- Different basic disciplinary knowledge and background;
- Basic skills for literature survey;
- Laboratory and project design;
- Scientific working knowledge;
- Mathematical skills;
- Documentation skills.

"Communication and management skills" are mentioned as a future success factor more frequently than "Basic engineering skills" (Table 8).

A comparison with topics and abstracts of EESD 2006 and SEFI 2001 conference proceedings shows that good and firm disciplinary knowledge and background are considered as general and fundamental future success factors for engineers.

The category "Cooperation skills" includes the following factors:
- Integration of partners (university, economy, politics);
- Ability to work with people from different backgrounds;
- Team work;
- Group work (theory and practice);
- Networking.

Table 8 shows that cooperation is considered an important future success factor for 11 out of 44 respondents. Respondents emphasised that different cooperation methods should be learned and practiced. A sub-group present in this case is a female group. 6 of 11 responses belong to this group of female engineers.

Networking, team work and other types of cooperation are also important general success factors addressed in the topics and paper abstracts of the EESD 2006 and SEFI 2001 conference proceedings. The desk research shows that the number of engineers who view cooperation and interdisciplinary team work as important engineering skills is increasing and women often belong to this group of engineers.[39] This category of factors is an indicator of the need for more than conventional basic engineering knowledge.

The category "IT skills" includes the following factor:

- Information technology (working knowledge of IT and its applications and security).

IT skills are considered an important future success factor by 9 out of 44 respondents according to Table 8. These respondents have different nationalities without any specific similarities in their cultural backgrounds. In section 4.6 (Table 9: responses to question 12) we observe that "IT skills" are not mentioned as an important issue in the context of sustainable development. None of the respondents who took part in EESD 2006 considered "IT skills" as specific engineering success factors for sustainable development either. The desk research however shows that computer or IT skills are important general skills for engineering work.

The "Cross-disciplinary work skills" is the next category of success factors indicating the need for new knowledge to perform engineering work. This category includes the following factors:

- Inter- and multi-disciplinary understanding;
- Problem-solving;
- Cross departmental discussions.

The first glance at the survey results in Table 8 shows that the factor group "Cross-disciplinary work skills" is considered less important than basic engineering skills. However, through a closer comparison between Tables 8 and 3, we find out that this factor is strongly emphasized by 39 out of 43 respondents for the future of technical education. The 11 responses in Table 8 show therefore a double emphasis on the significance of this success factor. Therefore "Cross-disciplinary work skills" is the most outstanding and selected general success factor for future engineers.

---

[39] COM 2006.

Tho desk research showed many examples of topics and abstracts within EESD 2006 and SEFI 2001 conference proceedings on success factors such as inter- and trans-disciplinary work.

Cross-disciplinary work is essential for the category "Reflexive knowledge" which includes the following factors:

- Knowledge on implication and impact of technology (e.g., on society, people, environment, etc.);
- Environmental and energy issues understanding;
- System thinking;
- Future-oriented attitude;
- Environmental impact assessment skills;
- Cost impact assessment skills.

Table 8 shows that although reflexive knowledge fields are well known today, only a very small group (8 out of 44 respondents) directly referred to them as a future success factor in their responses to question 11. This is most likely related to the opinion that impact assessment is a type of quality management that is not considered an engineering task at the moment. The desk research shows that the topics and paper abstracts addressed in EESD 2006 and SEFI 2001 conference proceedings include specific factors from this group.

The category "International experiences and language skills" includes the following factors:

- Staying abroad for one or two periods;
- Participation in international projects;
- Language skills (e.g. English);
- Knowledge of at least 3 languages as underlying skills for international experiences and language skills.

Table 8 shows that 7 out of 44 respondents stated factors related to international experiences and language skills as important success factors.[40] These responses should be analysed together with the answers to question 7 in Table 5. Those results show that 29 out of 43 respondents selected international experiences and language skills during education as strongly important future factors. The 7 specifications in question 11 seem therefore likely to be an emphasis on this success factor in addition to the broader understanding of engineering. International ex-

---

40 Nevertheless, barriers for international experiences are multidimensional; they are cultural, economic and political. Since international experiences are a main factor in understanding the global perspective of environmental problems and building international networks, technical universities should emphasis measures for improvement of international experiences.

perience is a factor strongly connected to education policies and political and cultural conditions of a country. The respondents who emphasised this factor in both questions 7 and 11 belong to very different countries. It can be concluded that despite specific possibilities for international experiences and language skills a general need exists in engineering work for such experiences.

A comparison between EESD 2006 and SEFI 2001 conference proceedings shows that language skills and experiences abroad are emphasised more clearly in 2006 in the EESD conference on Engineering Education in sustainable development than in the SEFI conference in 2001. Since no sub-group for EESD 2006 participants is identified in the survey, the author can only assume that the need for international experiences and language skills has increased over the past years independent of the context of sustainable development.

The category "Practical knowledge gained from learning by doing" includes the following factors:

- Focus more on real problems rather than text book ones;
- Hands-on experience (or more practice work);
- Problem- and project-based learning.

A small group of respondents (7 out of 44) considered hands-on experiences and working with real world problems during technical education as very important success factors for the future (Table 8). These topics are also found in the EESD 2006 and SEFI 2001 conference proceedings as problem- and project-based learning to have a stronger focus on real problems rather than text book ones.[41] This result should be seen in combination with responses to question 4 of the questionnaire in Table 2. The majority of respondents selected external learning processes as strongly important success factors. The additional answers to question 11 can be seen as a specification and strong commitment to responses to question 4. The off-campus experiences or learning by doing therefore appear to be important general success factors for engineering in parallel to "International experiences and language skills" and "Cross-disciplinary skills".

An important result is also the significance of "Practical knowledge gained from learning by doing" for both male and female engineers. Male and female engineers emphasise not only problem-oriented learning but also off-campus experiences. This reflects their opinion on the importance of practical knowledge in the future.

---

[41] It is however a challenge for technical universities to develop methods for problem-based learning within a short education time. International cooperation and collaborations with industry and local communities could be used to create more possibilities for these experiences. For problem-based learning see glossary.

The category "Capability of thinking and acting beyond the borders of the company on a global and local level" includes the following factors:

- Global vision;
- Capability to understand holistic views;
- See the big picture; Global economy and culture understanding;
- Independent research.

Table 8 shows that a small group of respondents view this factor as an important "future success factor" for engineering. At this point it is not evident which type of socio-political and socio-economic conditions the respondents considered for their answers. An emphasis on economic, social or environmental aspects cannot be recognised from these responses. They could be interpreted as complementary specifications of the general need for a broader understanding of the engineering profession. Alternatively, we could conclude that a small group of respondents have a holistic view of sustainable development in the future engineering context.

A comparison to topics and abstracts of EESD 2006 conference proceedings showed a focus on a holistic view of sustainability and global economic challenge. Issues such as climate change negotiation models and global and holistic pedagogies are examples of special issues (Cyrille 2006). The SEFI 2001 proceedings include topics on the global economy.

The category "Capability of critical thinking and responsibility" includes the following factors:

- Critical thinking;
- Social responsibility (e.g., in relation to people and environment);
- Ethics in technology and business and at work (e.g., in relation to biotech and genetics).

Since this factor is only mentioned two out of 44 times in answers to question 11, it does not appear at first glance to be a main general future success factor for the respondents (Table 8). In the next section it will be shown that "Capability of critical thinking and responsibility" is regarded as very important in the respondents' answers to question 12. This group of factors is important from precaution principle and risk aversion points of view and plays a substantial role in sustainable development. The difference in answers to question 11 and 12 can be interpreted as difference between mid- and long-term future scenarios or between future scenarios with or without the context of sustainable development.

The observation that these factors are specific to sustainable development is in line with the results of the desk research. A comparison with topics and abstracts of EESD 2006 shows that critical thinking, social and environmental responsibility are important factors as well as the need to raise awareness amongst students for the influence of social values on R&D. The SEFI 2001 proceeding does not clearly address these issues.

The category "miscellaneous factors" includes the following factors:
- Dealing with multiple issues at the same time;
- Open and broad minds;
- Analytical thinking;
- Problem-solving skills;
- How to interpret knowledge to practice;
- Creativity;
- Flexibility;
- Self confidence;
- Career development;
- Basic environmental knowledge.

These factors are mostly essential personal skills that are also identified in the EESD 2006 and SEFI 2001 conference proceedings. Most of these factors need to be addressed at school. They influence many other engineers' capabilities. Nevertheless, they can be improved or even affected during technical education. It is therefore important to discuss these factors in a broader context of education.

The analysis of responses to question 11 shows more similarities to the results of EESD 2006 compared to SEFI 2001 proceedings. This finding can also be explained by the development of engineering challenges between 2001 and 2006.

A more comprehensive analysis of the responses to question 11 in connection to results of questions 4 to 10 shows that the priority list for future success factors in responses to question 11 would change, if we take into account the entire responses of each questionnaire.

In the next section these results are analysed in the context of factors which have been important for the respondents' own work lives. This comparison is important not only for understanding the needs for change but also for a deeper understanding of relations between different factor categories. As mentioned earlier, these categories are developed for a better presentation of results of this analysis and should not be considered as mutually exclusive categories.

4.5.2 Differences between own experiences and perceived future success factors

Table 8 in the previous section and Table 10 in Appendix B both show that about half of the respondents chose one or two future success factors similar to those for their own success. Only a very small number of respondents, however, chose the same combination and priorities for such factors. The majority of respondents listed new essential success factors for the future.

From the respondents' point of view future focus lies on communication, presentation and management skills. The basic engineering knowledge was more important for the respondents' own career than for technical education in future where it was considered to be less important than cross-disciplinary work and non-technical knowledge.

A closer look at the individual responses in each category in Table 8 shows some interesting findings as follows:

Most respondents who regarded the "IT skills" category as their own success factor selected it as a future success factor as well. In addition the respondents indicated the need for different skills as a supplement to the basic engineering skills.

More than half of respondents who stated these factors as their own success factors did not mention them as future success skills (respondents [14], [17], [19], [25], [33], [35], [37], [41]). These respondents included a number of other categories for future success factors. Some examples are listed below:

- Respondents [19] and [37] show the factor group "IT skills".
- Respondents [17], [33] and [41] include "Management skills".
- Respondents [33], [35], [41] include success factors of the category "Practical knowledge by learning by doing".
- Respondents [19], [25] and [37] mention the importance of the category "Reflexive knowledge".

A group of respondents, who chose basic engineering skills both as their own and future success factors, follow an interesting trend, as well. The results show the importance of a broader engineering understanding in the future, even for the respondents who emphasised basic engineering skills as future success factors:

Only one of the respondents listed "Basic-engineering skills" and "IT skills" as future success factors (respondent [40]). Nine other questionnaires include different non-basic engineering factors such as "Communication and presentation skills", "Management skills", etc. in addition to basic engineering skills ([5], [6], [10], [11], [12], [28], [31], [34], [38]).

The category "International experiences and language skills" is mentioned by respondents with very different points of views. It seems to be important for maintaining a career in international companies, being able to take on projects in different countries or for understanding the engineers' global responsibilities. These skills seem like supportive activities and skills for basic and non-basic engineering skills.

"Practical knowledge by learning by doing" is a category with a different focus with respect to the past and future. Respondents who referred to these factors as their own success factors focused on research or teaching skills required taking theoretical and practical courses, while the future success factors in such categories directly imply more practice in research and working on real-world problems. The respondents who selected the category "Practical knowledge gained from learning by doing" as future success factors followed the idea of hands-on and more practical experiences during their technical education in order to be fit for economic and environmental challenges:

Respondents [5], [16], [31], [32] and [33] who included their own success factors "Practical knowledge by learning by doing" as research skills or hands-on practices and all mentioned "Management skills" as future success factors, too. Respondents [16] and [32] also referred to "Communication and presentation skills" as important future success factors. These responses can be interpreted as a need for change in the role of engineers from research to management staff or the need for additional management skills to be able to work in smaller and medium-sized firms without large research units.

The category "Capability of critical thinking and responsibility" noticeably was not present in responses to the question 10 and 11. However, the next section will show that the responses to question 12 are strongly related to this factor.

4.6 Respondents' suggestions regarding sustainable development (question 12)

In the last step respondents were asked to offer suggestions or describe their visions regarding engineering and engineering education in sustainable development in 2020. 2020 was chosen to create a link to some contemporary engineering discussions on engineering visions.[42]

The Respondents' regarding future success factors within sustainable development (question 12) are documented in Appendix B to allow the reader to make his/her own judgment.

These responses were analysed and grouped into the same factor types used in Table 8 for question 11. This categorisation allows for comparison between responses based on respondents' individual assumptions in question 11 and their visions of the future in context of sustainable development in question 12.

---

[42] Such as (NAE 2005).

Table 9: Categorization of issues on the future visions suggested by respondents in the survey [43] (question 12: Suggestions with regard to engineering for a sustainable development in the future (2020), see Appendix B)

Factor Type	Code
Cross-disciplinary work skills	[1], [12], [13], [16], [19], [21], [32], [36], [37], [38], [39],[41], [43]
Capability of thinking and acting beyond the borders of the company on a global and local level	[4], [13], [14], [15], [16], [19], [21], [31], [36], [39], [43]
Capability of critical thinking and responsibility	[7], [9], [10], [11], [13], [14], [15], [21], [39], [41], [43]
Basic engineering skills and new research skills	[1], [5], [10], [32], [34], [37], [42], [44], [19]
Cooperation skills	[2], [17], [34], [35], [41], [42]
Practical knowledge gained from learning by doing	[6], [7], [8], [34], [41], [42]
Communication and presentation skills	[12], [21], [31], [34]
International experiences and language skills	[2], [8], [34]
Management skills	[11], [33]
Reflexive knowledge	[5], [36]
Miscellaneous factors	[19], [22], [31], [21], [37], [38]

Responses in Table 9 show a different observation from Table 8. Table 9 relates to question 12 where respondents were asked to think in the context of sustainable development. The analysis of their responses shows an understanding of the holistic view of sustainability that is also important for the technical dimension of sustainability. IT skills a factor often mentioned as a future success factors in Table 8 is not present in Table 9. This finding can be interpreted as "IT skills" are not important for sustainable development or they will be a part of basic engineering in the future.

Cross-disciplinary and trans-disciplinary work skills are considered to play an underlying role for successful engineering work in the future.

Capability of thinking and acting beyond the borders of the company on global and local levels are two main principles for sustainable development. These factors seem to be as important as basic engineering skills in the context of sustainable development.

The higher number of responses in relation to the "Capability of critical thinking and responsibility" factor group in Table 9 is an indicator of awareness of the precautionary principle that is a basic requirement of sustainable development.

---

43 Interpretation and categorization of information was assisted by Dr. Christina Raab.

This factor has been mentioned at least once by almost all nationalities present in the survey group.

Responses to the survey question 12 leads to the following results:

- different understandings of sustainable development;
- strong need for non-basic engineering skills;
- manifold needs for changes of curricula of technical education in a sustainable development.

These issues will be addressed in the following sections.

4.6.1 Different understandings regarding sustainability

The respondents' statements show that different understandings and interpretations for sustainable development exist within the group.

Some answers from the USA emphasised the main role of system optimisation for economic growth for sustainability (A). There were respondents from the USA and Europe who focused on the engineering profession's global responsibility. One engineer used both types of arguments (B). Some respondents from Europe focused on the ethics with a wider vision (C).

Mode A: Focused on the system optimisation and economic growth: see respondents: [16], [36]

Mode B: More global responsibility was mentioned in questionnaires: [15], [21], [36]
Some respondents from Europe added suggestions to this group for more emphasis on the society's needs: see questionnaires: [7], [9], [35], [38], [41], [43]

Mode C: Responses with messages on wider vision, holistic view and ethics were given by interviewees who took part in the EESD 2006 Lyon: see questionnaires: [4], [39], [43]

The explorative character of the survey only allows for the conclusion that there are different understandings of sustainability and to assume that more interpretations exist than identified in this study.

Cultural differences are visible in results. Nevertheless it is a separate research question, whether responses can be generalised or if they depend on respondents' individual cultures and experiences.

4.6.2 Strong need for non-basic engineering skills

Table 9 shows that non-basic engineering skills are strongly important for the success of engineers in the future. The respondents' individual statements in Table 11 (Appendix B) show different backgrounds for this opinion. Two backgrounds that are obviously present are the challenges of solving complex problems by holistic views and the challenge of career development.

Responses in modes B and C described in section 4.4.1 are an indicator of the need for new non-basic engineering skills to overcome the challenge of social responsibility of engineers and the need for a holistic view.

> "Universities should advocate multi-disciplinary learning and teach global issues related to health, environment, culture and economy."
>
> A colleague from USA (see Table 11, questionnaire: [15])

A European participant of EESD 2006 sees a holistic view as an essential necessity for engineers:

> "In 2020 engineering education must be able to guide students to even more holistic comprehension of the global phenomena as today. Today's engineering students are taught details. In 2020 the fractioned knowledge "feeding" necessity will have become outdated, whereas different aspects are always taken into account with appreciation"
>
> (see Appendix B, Table 11, questionnaire [4])

The importance of "Management skills" and leadership skills for career development is mentioned by a colleague from the USA according to his personal experiences:

> "One of the major areas that the technical curriculum has missed out on is 'Leadership'. They don't try to prepare their students to become a leader. They try to focus so much on the technical aspect that they completely ignore the other aspects. ...Most of the companies' top executives have MBA degrees, which send a message that technical people are not as important ... I have seen people with little technical background who could give effective talks and presentations in meetings and company gatherings have excelled in their career very quickly and moved to the top of their professional path."
>
> (see Appendix B, Table 11: questionnaire [32])

Not only leadership but also other social skills were regarded as necessary for career development. A colleague from Sweden described the importance of building networks as follows:

> "It will probably also be more important to create the right networks (i.e. get to know people from business and for international work) already during studies."
>
> (see Appendix B, Table 11: questionnaire [42])

These respondents state that engineers who work in large industrial organisations should be ready for leadership and those who work in small companies should be able to network successfully. According to these statements technical education is responsible for the practical preparation of engineers for these challenges.

Such capabilities of engineers lie indeed beyond the contemporary basic engineering skills.

4.6.3 Needs for changing curricula

All interviewees made suggestions regarding curricula changes. Some selected key suggestions are described below:

- the need for balanced curricula to gain practical and theoretical knowledge: see Appendix B, Table 11: questionnaires [1], [7], [6], [10], [11], [13], [17] [34], [39]. This point has been discussed previously (Chapter 3) through the discussion about the tension between science and engineering;
- the need for knowledge of best and worst practice of engineering is mentioned in questionnaire [36]. It could be interpreted as a means to learn reflexive and critical thinking during technical education;
- the need to motivate students is mentioned here for the first time: see Appendix B, Table 11: questionnaire [35]. Motivation of students implies that engineering education provides settings and perspectives that are attractive for young people. At the same time motivation has its roots in the world outside the university. School children have already a perceived idea about the role of engineering based on the information provided to them and their experiences. Career opportunities, social understanding of engineering, individual opinion of friends and family and many other factors influence the motivation for engineering students as well. Therefore such statements can be interpreted as a demand for a new understanding of the role of engineering profession in the societies.

4.7 Future opinions with and without focus on sustainability

The focus on different factor groups was very different for question 11 on future success factors and question 12 with specific focus on the context of sustainable development.

Factors such as capability of critical thinking, taking responsibility, and thinking and acting beyond the borders of the company at global and local levels were considered more important in responses to question 12 than in question 11. This result shows that respondents differentiated between the general and sustainability-specific success factors.

For question 12 the respondents assumed socio-economic conditions and environmental situations demanding more critical thinking, responsibility and a holistic view of engineers.

The responses to question 11 could be interpreted as general success factors for future engineers or they can be viewed within the socio-economic context that differs from sustainable development.

In other words the difference between responses to question 11 and 12 could be understood as

- differences between general and sustainability specific success factors or
- differences due to two different social values for technology development and engineering profession.

In both cases these results show the implicit assumption by the respondents on the *interrelation of technology development, engineering profession and engineering education with social values*. This interrelation exists through social values and socio-economic and socio-political conditions in different societies.

4.8 Conclusion

This chapter has shown divergent opinions of a small group (44 respondents who are mostly engineers), with various backgrounds and nationalities, regarding success factors for engineers and their suggestions for technical education in 2020 in the context of sustainable development. The respondents included both international experts with a wide spectrum of activities as well as engineers engaged in core engineering tasks.

Future success factors such as critical thinking and responsibility, thinking and acting beyond the borders of the company at global and local levels were identified.

Results of the survey (especially responds to questions 10, 11 and 12) show three main aggregated factor groups, namely

- Conventional basic engineering skills;
- New basic engineering skills;
- Non-basic engineering skills.

The following categories are viewed as the new basic engineering skills:

- "Communication and presentation skills";
- "Management skills";
- "IT skills";
- "International experiences and language skills".

Various non-basic engineering skills and knowledge factors could not be distinctly categorized. They can be found within the following main categories and are considered as additional skills to "Basic engineering skills".

- Personal skills (in the "Miscellaneous" category and in "Capability of critical thinking");
- "Practical knowledge gained from learning by doing";
- "Reflexive knowledge" and "Capability of thinking and acting beyond the borders of the company".

The above list of knowledge and skills for individual engineers has a strong relationship with the following list of skills and capabilities:

- "Cooperation skills";
- Cross-disciplinary working skills.

Especially in the context of sustainable development there is a tendency for the following non-basic engineering skills for individual engineers and technical groups:

- "Capability of critical thinking and responsibility";
- "Capability of thinking and acting beyond the borders of the company".

The results on different categories show the interdependency of the defined categories only in the explorative survey and should not be over-interpreted. The direct result of the survey, to be used for further analysis is the presentation of the diversity of terms selected by engineers with different backgrounds and their responses to questions about the important skills in the past and present as well as future success factors for a career as an engineer.

4.8.1 Some important details

The results are used as a new information source in engineers' language for a qualitative comparison with results available from a literature survey in order to show new challenges for engineering.[44]

The majority of the respondents in this survey considered teaching experience less important than other factors.

---

44 The survey was performed with a questionnaire that was communicated to respondents via e-mail or telephone calls. The questionnaire consisted of three parts. The first part of the questionnaire included questions 1 to 3 documenting the background of each respondent. The second part of the questionnaire (Q4-Q9) with questions on success factors designed to identify the respondents' opinions on the importance of some key factors in the Bologna process and the Barcelona declaration. Respondents were also asked to comment on their own experiences with respect to these factors.

Some engineers in the survey more strongly emphasised structural changes for a broader engineering concept. All of them took internationalisation as an initiator for reforms but a few considered local experiences and engagement as important as international orientation. Research and learning by doing are given a higher priority than teaching experience during education.

In the third part of the questionnaire (Q10-Q11) respondents had the opportunity to give success factors that have proven important for them in their own career (Q10) as well as for future engineers (Q11).

Relative to their own success factors the majority of respondents referred to (entirely or partly) the new factors as "future success factors". Examples of such new factors are basic entrepreneurial skills, rhetorical skills, participation in international projects, capacities to understand holistic views, career planning or collaboration skills. These stated factors are categorized into 11 groups. These categories are sorted based on the respondents' assigned priorities as follows:
- "Basic engineering skills";
- "Communication and presentation skills";
- "Management skills";
- "Cooperation skills";
- "IT skills";
- "Practical knowledge gained from learning by doing";
- "Cross-disciplinary work skills";
- "Reflexive knowledge";
- "International experiences and language skills";
- "Capability of thinking and acting beyond the borders of the company at global and local levels";
- "Capability of critical thinking and taking responsibility".

Miscellaneous factors included: Analytical and logical thinking; Problem-solving skills; Persistence/endurance; Independency; Flexibility; Self-confidence; Concentration; Life-long learning; Learning skills; Feedback; Patience.

With respect to the "future success factors" priority the list of categories is sorted as follows:
- "Communication and presentation skills";
- "Management skills";
- "Basic engineering skills";
- "Cooperation skills";
- "IT skills";
- "Cross-disciplinary work skills";
- "Reflexive knowledge";
- "International experiences and language skills";

- "Practical knowledge gained from learning by doing";
- "Capability of thinking and acting beyond the borders of the company on a global and local level";
- "Capability of critical thinking and responsibility".

Miscellaneous factors included in this case are: Dealing with multiple issues at the same time; Open and broad minds; Analytical thinking; Problem-solving skills; How to interpret knowledge to practice; Creativity; Self confidence; Career development; Basic environmental knowledge.

Responses to question 11 represent the respondents' opinions on trends in the future requirements for engineering capabilities from their own points of view. A comparison of responses from European and American respondents (each accounting for about half of the group) shows that inter- and trans-disciplinary work and understanding are regarded in both groups as important factors. International projects and global issues, presentation skills, team work skills, critical thinking and responsibility are other factor groups which are common in European and American responses.

European respondents emphasised the importance of knowing more than one foreign language. This is probably due to the need of European engineers for co-operation throughout the European Union in different languages. In international research English language skills plays a significant role and are regarded as very important. Another difference is the stronger focus of respondents in the USA on IT skills as a future success factor. A comparison of this finding with the discussions among engineers in an open Internet forum (Appendix A) leads to the observation that engineers in the USA perceive developing and using IT skills as a higher challenge compared to engineers in Europe. This is an indicator for different challenges for the future of engineering in different countries.

Other findings related to responses from the entire respondents group to question 11 are as follows:

Most factors selected by the respondents to question 11 are non-basic engineering factors. These capabilities are all required for engineers who work outside industrial organizations. The system of preparing engineers primarily for industrial work is perceived as outdated and not applicable to future needs. The factors, however, are not typical for engineers' entrepreneurial capabilities and their own development of new technical solutions, because of the emphasis on cooperation. The respondents' opinion on this matter reflects the need to prepare engineers for an intra-entrepreneur engineering system. The success seems to be tightly combined with non-basic engineering skills and practical knowledge. This observation is based on the following results:

"Communication and presentation skills" and "Management skills" are mentioned much more often than other factors. Basic engineering skills are stated as often as cooperation and cross-disciplinary work skills. Analysis of the responses to questions 4 to 9 shows other priorities as well. "Cross-disciplinary work skills", "International experiences and language skills" and hands-on projects are in general more often mentioned than "Communication and presentation skills". This can be seen as evidence for the importance of change in technical education to prepare future engineers through cooperation of engineers with other professions. The significance of international experiences and language skills can also be understood in this context. The globalisation and internationalisation of technology development requires engineers who can deal with international cooperation. Engineering students should be prepared for this challenge.

Although only a smaller group of respondents explicitly viewed learning by doing as an important success factor, most of them indicated the importance of experiences through hand-on and local or global projects during their education. In pedagogic literature, learning by doing is considered as a core element of special educations fields such as engineering education. This type of learning requires cooperation between technical universities, industry and local communities and can be supported by international cooperation as well. It therefore implies engineers developing both inter- and trans-disciplinary skills.

The importance of "Practical knowledge gained from learning by doing" can also be explained by the interdependency of success factors for creative engineering work. The success of radical technical innovations in the real world depends often on the integration of the development in local, regional and global learning networks.

The challenges for engineers in this regard are to:
- be individually creative;
- work in trans-disciplinary teams;
- be able to understand systems;
- consider the dynamics and changes of stakeholders' interests during a rather long development phase of technology.

The list above shows the integrative character of challenges as a reason for the importance of problem-based learning. In this case students cannot learn different skills while isolated from one another. They can gain experiences only through involvement in projects and allowing them to analyse success and failure factors for their projects.[45]

---

45 The dilemma with these examples is, however, the lack of such projects throughout education. Problem-based learning is possible, when the technical university has a chance to take part in real technology research and development projects. To conduct such projects at technical universities or at technology and science parks of universities, universities need financial sources for such training projects (see also Albert et al. 2001).

The fourth part of the questionnaire (Q 12) contained a question asking for respondents' suggestions for future engineering in the context of sustainable development in 2020. The survey showed different understandings of sustainable development among the respondents. Sustainability was understood as a context for a holistic view and engineers being more responsible and responsive to social needs, and to focus on the global economy when optimizing the economic growth.

The analysis of statements and extraction of factors related to the 11 factor groups shows that the following required capabilities are mentioned more often than other factor groups:

- "Critical thinking and responsibility";
- "Capability of thinking and acting beyond the borders of the company on a global and local level" and
- "Cross-disciplinary work".

The difference between responses to the last question on future success factors without a special context of sustainability showed that for question 11 respondents used a different priority compared to question 12. The difference between the responses to questions 11 and 12 could therefore be an indicator for respondents' opinion on the influence of political, social and economic condition on engineering profession and engineering education.

At the same time the respondents' answers to question 12 can be considered as special success factors for engineering in the context of general sustainability principles. The reflexive knowledge and critical thinking factor group are the underlying factors for thinking and acting beyond the borders of the company and acting according to precautionary principle, local engagement, global responsibility and fairness.

Other respondents focused on today's core engineering activities and recommended modernisation of education programs with updated content. These are some of the similarities and differences of opinions within a small group of respondents (mostly engineers) which show the need for changes based on different interests and values.

"Capability of thinking and acting beyond the borders of the company" presents a holistic view and the need to understand different action levels. It is possible to understand the global dimension of risks which are caused by technical solutions. As mentioned earlier we can assume that some respondents perceive that in the future there will be no other choice than integrating a holistic view of sustainable development into engineering work. In other words, these respondents consider such special factors for sustainable development (considering social, environmental and economic aspects, etc.) as general success factors for future engineers. Nevertheless at this point the reasons for higher significance of the factor in question 12 and the big difference between priorities mentioned in question 11 and 12 cannot be analysed deeply enough and needs further analysis through future research work.

One of the main results of the survey was the need for responsibility of engineers in the context of sustainable development. This is an important issue that influences the role of engineering in the society and should be studied for the analysis of changes in technical education. It will therefore be discussed in more detail in the next section.

4.8.2 Demand for a higher responsibility of engineers

The need for individual engineers' higher level of responsibility of and support of this responsibility by their communities is mentioned in different forms in the survey responses. One example is presented below:

> "An economic world would be desirable, where engineers could refuse any 'not honourable' work. It means that it would not be socially 'acceptable' that an engineer works for a highly polluting enterprise."
>
> see Appendix B, question 12, questionnaire [39]

Individual engineers' responsibilities and adaptation of holistic views to understand impacts of technology application changes the practice in engineering. It implies a new definition of the interaction among design and production, design and programming, design and building, etc.

According to the model presented by Bill Buxton for software design (which could be applicable for many other engineering fields as well), design task[46] presents the starting point with the main responsibility of designers at the beginning and a gradual increase of the responsibility for other engineering practices, business management and sales. During the entire product life cycle designers remain in charge of the entire process and take part in decision-making on further development of products, processes and services. They share the responsibility with other groups at different phases of development. Their responsibilities will primarily be defined by social, cultural and environmental preconditions and partly by quantified standards and norms of products and processes.

One of the respondents demanded a general obligation of holistic comprehension as follows:

> "An engineer should be more capable to cope with the full process in which he/she is involved and have more holistic comprehension of today's global phenomenon."
>
> see Appendix B, question 12, questionnaire [39]

The concept of engineers' responsibility is presently applicable only for a small number of engineers who can be involved in the decision-making process during

---

46 See (Purgathofer 2006).

all phases of the product life cycle. One of the issues with this model is that it does not directly address the issue of the fragmentation of engineering activities. The structure of large industrial R&D departments generates a fragmented picture of the technology for the engineers involved with the related projects. Responsibility for the engineers' own design towards society is transformed into responsibility towards the company. Organisational changes could support in this case the behavioural changes[47].

Bill Buxton's model also has other limits: Responsibility for the whole process could overstress individual engineers in large projects. An indicator of this problem is the biography of engineers who committed suicide because of their failures or presumptions that their design could fail.

Some relevant issues that should be discussed between engineers and society are as follows.

Questions related to needs for organisational and behavioural changes for responsibility are as follows:

> What is the difference between individual responsibility of engineers and their responsibility in their organisations?

> Which individual capabilities and institutional structures are necessary to deal with the responsibility of engineering?

Questions related to needs for conceptual changes are as follows:

> Which responsibility models are appropriate for which engineering field and task?

> How should responsibility be learned?

> Should engineers learn to deal with responsibility for the impacts of technical solutions during their practice upon graduation or should they learn by doing during their education?

The list of issues for discussion will grow, as we specify different responsibilities for engineers. A concept for engineers' responsibility should make up a balance between all three categories of organizational, contextual and behavioural challenges to address their individual, group and social responsibilities.

Part II of this book investigates challenges for the engineering profession in the context of decision-makers during technology development and application. The analysis begins with elaboration of a concept for a socio-technical system.

---

47   A result of desk research: identified categories of needs for engineering proficiency in chapter 3: organizational, conceptual and behavioral changes.

Part II

Technology – Engineering
Challenges according to technology development and application

Part I showed arguments for the influence of social values on the engineering profession. The main purpose of this part is to provide a brief overview of the influence of social values on the decision-making mechanism of technology development and applications that determine an engineers' work.

5. Context of technology development and application

Challenges to engineering work in the future were discussed in the previous part of the book based on the engineers' own opinions, declarations of the engineering associations as well as references on engineer's role in society. In this part such challenges will be investigated in a broader context of technology development.

Technology development and application imply the involvement of many different actors and professions with different intensity and scope of involvement and responsibilities. People with different interests, knowledge, skills and background operate individually, in small groups or in international organizations in diverse activities such as resource extraction and preparation, process development, product manufacturing, marketing, etc. These activities are influenced by political, financial and economic conditions in a society. Engineering work is one of the key activities in this context.

Therefore in order to investigate challenges to engineering work, we also need to understand determinants of technology development and application.

For the investigation of these determinants the author used sources related to technology development and application in the context of innovation, social or environmental policy. In addition a number of lectures and conference proceedings that show different descriptions for terms "technique" and "technology" have been used. In references that are cited in this part of the book technology is related often to the value-oriented terms such as growth, development, efficiency, productivity, innovation, sustainable development, power or crafts, risks, security, fairness, life, impact, etc.

A group of authors such as Jischa (2004) focus on the central role of technology in different cultures and epochs. Jischa starts his analysis with the Neolithic period and the development of agricultural techniques. In his book he presents an analysis of the development and the role of technique. His analysis includes some aspects of the interaction of technology development with social, cultural and political developments and the influences of wars, people's immigration, etc. on technology development. He describes developments as a result of human activities (economic, political, scientific, and social) based on interactions between resources (material, energy, soil, nature and knowledge), visions (values and norms, religion, ideology, science) and institutions (infrastructure, informal structures, policy, and regulation).

An older definition by Gendron emphasises (in 1977) practical and theoretical knowledge which are embodied in "productive skills, organization, or machinery" by technology development (Porter et al. 1980, P. 11).

The desk research shows that the influence of social context on technology development has a central role also in critical theories of technology[48] that claim that technologies embody the values of their development context.

> "Modern technology embodies the values of a particular industrial civilization and especially of its elites, which rest their claims to hegemony on technical mastery. We must articulate and judge these values in a cultural critique of technology. By doing so, we can begin to grasp the outlines of another possible industrial civilization based on other values."
>
> (Feenberg 1991; Borgmann 2006)

Identification of different existing social values as the background of industrial civilization indeed requires a deep and comprehensive analysis of interactions between social and technological development. In this chapter the author presents a brief overview of some examples of the contexts of technology development. Chapter 3 "History of challenges for engineering" included results of a research on the history of engineering and presented literature from different disciplines such as history (Akin 1977; Bellis 2007; Canel 2000; Canel et al. 2000; Hill 1996; Hill 2005; Kaiser/König 2006; Moon 2005; Rae/Volti 1999; Rolt 1958), references from technical museums and media (Cornish Mining World Heritage 2007; Spartacus educational 1999; Technology Museum of Thessaloniki 2001; The National Museum of Science and Technology in Stockholm 2006; The Times 2001), a number of scientific references on society, environment and technology (Blanc 2007; Dodds/Venables 2005; Jischa 2004; Jucker 1998; McCarthy 1989; Veak 2006; Kiper/Schütte 1998), EU research reports and official documents (COM 2006; COM 2004) and last but not least references with a special focus on engineering (ANSB 2005; Oldenziel 2000; Roman 2003; RAE 2003; Wulf 1998).

The results of this research show four different categories of engineering contexts for technology development and application:

- elite-oriented context;
- science-oriented context;
- need-oriented context;
- social process-context.

These four contexts are not fully independent of one another. The elite-oriented form has existed from ancient times up until today. Examples in this chapter will point out that different social values have been generated sequentially without eliminating previous values. Today all of these contexts can be found to co-exist with each other. The next sections show that challenges to the engineering task are context-specific.

---

48 Glossary.

5.1 Elite-oriented technology

Technology is considered from an elite view point as extremely positive because it fulfils the needs of human beings and is interpreted as a means to an easier life: "Technology includes the use of materials, tools, techniques and sources of power to make life easier or more pleasant and work more productive."[49]

The inclusion of the target to make life easier and more productive is combined with the importance of technology in society's development. The role of tools that made life easier or surviving possible was one of the most important human developments in ancient times. It is therefore easy to understand the elite-oriented context of technology that we read about in ancient myths.

In addition to the fulfilment of needs the power component is explicitly addressed in the above definition. It has also been one of the essential needs of society to serve and defend itself by means of better tools and techniques. Box 4 gives examples that show the elite-oriented context of technology. It originates from the fulfilment of essential human needs and its contribution to social development by way of access to power over nature and defence against dangerous threats from the outside through the use of different techniques and tools.

Box 4

Elite and technique in history

In oriental history during the Sumerian, Assyrian and Persian times from four thousand years before Christi to the birth of Christi, technique was introduced by kings who saw themselves as God's representatives.

In Persian myths the control over nature by knowledge and wisdom was personified in the elite class. Myths describe the Shahs (kings) as inventors and introducers of the new ways of life.[50] Hakim Abol-Ghasem Ferdowsi Toosi the world famous Persian (Iranian) poet (940-1020 AC) tells the following hero tales in a book named Shahnameh.

"Kaiumers first sat upon the throne of Persia, and was master of the world. He took up his abode in the mountains, and clad himself and his people in tiger-skins, and from him sprang all kindly nurture and the arts of clothing, till then unknown."

"Husheng first gave to men fire, and showed them how to draw it from out the stone; and he taught them how they might lead the rivers, that they should water the land and make it fertile; and he bade them till and reap. Tahumers, his son, was not unworthy of his sire. He too opened the eyes of men, and they learned to spin and to weave ..."

---

49 Http://www.britannica.com.
50 Hakim Abol-Ghasem Ferdowsi Toosi the world famous Persian (Iranian) poet (940-1020 AC) tells these hero tales in a book named Shahnameh The Shahnameh or The Epic of Kings is one of the definite classics of the world. It tells hero tales of ancient Persia. The contents and the poet's style in describing the events takes the reader back to the ancient times and makes he/she sense and feel the events. Ferdowsi worked for thirty years to finish this masterpiece." (Zimmern 2001).

"Jemshid first parcelled out men into classes; priests, warriors, artificers, ... And the year also he divided into periods. And Persepolis was built by him that to this day is called Tukht-e-Jemsheed, which being interpreted meaneth the throne of Jemshid..."

Starting in antique times there are references to individuals with knowledge to design and use instruments for construction, planning, transport, etc. These people were not kings or lords but they represented the elite class of society.

The use of technology was directly connected to military power and political motivations. The Trojan horse myth shows the key role of technology in the military success in myths.

Greek myths include clear examples of the elite role of technology for society by the invention of important cultural artefacts. Humans needed in these myths to steal these techniques or learn them from Gods:

The titan Prometheus the friend of humans stole fire from the gods and gave it to the mans[51]; the king Tantalus stole nectar and ambrosia from Zeus' table and gave it to his own people[52]; revealing to them the secrets of the gods; Demeter, the goodness of grain and agriculture, taught humans agriculture[53].

The dominant role of technology, its elite role and the importance of tools and techniques for social developments are present even in our present language:

> "Our history is filled with records of our tools and technologies. Epochs are measured (today) by their most important technological developments. The Stone Age is followed by the Bronze Age which is followed by the Iron Age which is followed by the Steel Age."[54]

The elite role of technology can be found in cases that because of certain reasons technological development has a very strong impact on cultural rules.

Is the application of a certain technology always combined with more social benefits?

The shadowy side of an elite-oriented context appears when it is associated with an element of political or military power to pursue certain technologies without a democratic and human friendly critical selection mechanism. Engineers should decide in such cases between their responsibility to consider potential negative impacts of technologies, and their wish to have a privileged role in an elite-oriented context for technology development.

5.2 Technology as an application of science, a new elite role

A different picture of engineering work emerges, if technology is interpreted strongly as the use of science. In this case it is the science that drives the technology development further. As long as science is developed, technology could

---

[51] (Encyclopedia Mythica 2001a).
[52] (Encyclopedia Mythica 2001b).
[53] www.waltm.net/demeter.htm.
[54] www.regent.edu/acad/schcom/rojc/mdic/history.html.

be developed further and be used whenever appropriate conditions are available. The seed is prepared by science and it will grow, if it finds appropriate conditions, argues this perspective. This interpretation also implies a high degree of freedom, free from local context, for technology research, development and application. A difference to the elite role of technology is that this time not the kings, mythical gods or political powers play or demand this elite role, but the science claims this role.

This idea is presented in the works by Bacon:

Box 5

Elite role of technology through science

"Francis Bacon (1561-1626), natural philosopher, saw science and technology as means to understand and master nature. Bacon believed that the magnetic compass, the printing press, and gunpowder were the most important developments of modern man.

Bacon's contemporaries Rene Descartes (1596-1650), G. W. Leibniz (1646-1716), and Blaise Pascal (1623-1662) saw the world as being controlled by mathematical principles. For the fathers of rationalism, technology and science were tools to understand and master the world. Descartes's exclamation, "give me matter and motion and I will construct the world" pointed to his mechanistic worldview..."[55]

This interpretation of technology is still present in certain fields. Biotechnology is often seen as science-driven technology, which helps human beings to understand the origin of life. A stronger link exists between freedom and the elite role of science and technology in nuclear technology. The development of this technology was not possible outside scientific laboratories by autodidact scientists. The creativity of engineering is in this case a necessity, but not the elementary component of technology development. Nuclear power was developed with the help of the physics of atoms and radioactivity with the political motivation of the Second World War.

Scientists with a technical background, such as Wilhelm Roentgen, who were experimentalists in the early years of the twentieth century, compensated the lack of modern facilities at laboratories with their scientific and theoretical work. The trial and error experiments to arrive at inventions such as application of X-rays were however unavoidable. This example shows that engineers who work in a science-oriented context of technology development need a very strong theoretical background.

The history of nuclear technology and the risks connected with its military and civil applications reflect the risk of the elite role of technology through the elite role of science. In this particular case, the engineers have to be aware of the potential impacts of their work on society, although their working environment

---

55 (Regent University's School of Communication and the Arts 1997).

might be isolated laboratories. It is also possible that engineers have the capacity to issue early warning if they identify potential risks of their technical solutions.

5.3 Need-oriented context of technology

Some authors differentiate between science-driven and market-pioneering high-tech sectors (Lehner 2005) or science versus customer and demand-driven technologies. In this context technology is viewed as a means to fulfil customers' demands by technology based on science. The technology can be developed to fulfil special needs with the help of scientific knowledge. The engineer would be a designer who supports the interaction between needs and resources with his/her ideas and knowledge.

Information technology is an example that has been developed in this context. It includes scientific research by physicists, mathematicians and technical knowledge of electronic engineers. The technology development and application is based on the needs of scientists for more effective calculators, using the high potential of science and engineering. At the same time strong tensions between its generators occurred. The development and application of information technology has not always led to the fulfilment of user needs. The scientists therefore had to align themselves to the existing limited technical solutions and accept simple solutions which were useful for users instead of the best scientifically possible solutions.

New customer needs, which were generated after the mass distribution of information technology in the 1980s, were also only partly fulfilled. The growth of technology itself led to new needs and new customer demands. Impacts of information technology on daily life form a third additional factor to science and customer needs, namely the economic factor which has changed technology into a business tool. The interest of the economy in information technology might be one of the main reasons for the magic development of technology in less than a century. This example shows that a need-oriented context of technology with different possible motivations and social values (fulfilment of customer needs, or scientific improvement) may change more easily during the development of technologies. New social values can be integrated with the original motivations and change the development path of technologies.

Engineers need to consider in this case different interests related to market needs. The engineer stands this time not as elite above society's judgment. Under these circumstances an engineer is like a manager operating under clear orders by the shareholders who can be society or individual investors. The engineer has in this context a more direct dependency on the user. It is also possible to see this dependency in a broader context on society's needs and judgment. Engineers are expected in the latter case to have a good understanding of the benefits and risks of technologies not only for their company but also for society.

According to this context technology is not limited to artefacts and art that is developed and used for a special purpose; it is emphasised in the social sciences as an interactive social process for a special purpose.

> "... as Merritt Roe Smith has put it: technologies can and do have "social impacts," but they are simultaneously social products which embody power relationships and social goals and structures (Smith 1985). Social impacts and social production of artefacts in practice occur in a tightly knit cycle."
>
> (Edwards/Arbor 1994)

Balabanian (2006) describes the role of social processes in technology as the binding factor of elements of technology. The idea of socio-technical systems is clearly presented in his definition:

> "Technology is not simply a collection of machines, but the relationships among them, and the relationships between them and people. Just as a collection of words cannot adequately represent the rich texture of language, so also a collection of machines, even interconnected machines, cannot adequately represent contemporary technology."

If we accept technology as a social process, then we have to accept reflection of the man-made technology on the human being itself. A social process context does not weaken the role of science and creative technical abilities; it brings new actors and new rules to the arena. Socio-technical systems include stakeholders who represent people who are involved in the process and shape the technology development through their interest and value-laden activities. Development of the socio-technical generates impulses for social development but it is embedded in the social development itself. This interpretation of technology is compatible with the gradual shift in the moral customs and perspectives in interaction with technological development. Technology has at least as a decision-factor emerged to the policy and political practice of societies, and has changed from simpler political methods of the ancient Greeks to a more sophisticated political practice of modern state that embody technological ethos (Albin 2006). Due to this interpretation technological development has been one parameter in the change of human beings' sensibility towards nature. This interpretation will be used in the next section to develop a concept that considers social values and actors as shaping factors during the entire process of technology development and application.

5.4 Conclusion

This chapter presented four different understandings of technology development that show the interrelation between technology and its context and social development. The first three contexts focus on political elite-role, scientific elite-role and market needs, while the last context shows a more comprehensive interpretation of interactions between technology and social development.

The later context defines a socio-technical system that integrates engineering work as one of its activities and opens up the discussion on a broad spectrum of questions related to the communication between different actors and engineers and the responsibility of engineers during technology development and application. Some of these questions are:

How do engineers deal with social obligations outside the borders of the company?

Are engineers able to deal with social activities which go beyond technical factors?

How should technical education emphasise skills for communication, coordination and public participation processes to prepare engineers for their integrated role in the socio-technical system?

These questions can only be answered in relation to the definition of the role of engineering and technology development in the 21st century.

The interdependency of society and technology and the shaping of technologies are however very complex and beyond the scope of this book. A concept for a socio-technical system is presented in the next chapter to give an overview of determinants of technology development in a socio-technical system. Some highly relevant challenges to the engineering work can be demonstrated through analysis of this concept.

6. A concept for a socio-technical system

In the previous chapter a social process was defined as a socio-technical framework for technology development and application. Such a system combines technical development with the social and cultural backgrounds. This is an appropriate framework for the analysis of determinants of technology development and application in sustainable development. Rosen (2002) presents a definition for a socio-technical system that includes interrelations of *technological, social and cultural elements*: *Social actors* are designers, engineers, manufacturers, various groups of users and non-users, promoters and policy makers. *Technology* is specified by artefacts/products and their components, production equipment, technical knowledge and practices, (goals, problem-solving strategies, theories) and related artefacts such as different products in the same industry. *Culture* element includes cultural practices and narratives, cultural products, cultural activities such as gatherings and events, organisational cultures, publications, advertisements and other literatures, broader cultural values and resources. A common goal among socio-technical systems is using resources to fulfil perceived needs and market demands. *Resources*[56] are either human-related resources such as different types of knowledge, power, art, or physical resources, such as material, different sources of energy, documented information, data, and financial resources as well as time, machines or animal power.

Real needs of manufacturing companies are:

- solution of special problems within a sector (Transport, industry, agriculture, household, energy, construction, etc.);
- improvement of the technical efficiency after the development of the first generation of a given technology;
- improvement in the economic benefits to a firm, a group of actors or an entire sector.

Examples of additional national or international needs and requirements that imply application or development of technical solutions are:

- risk reduction (technical, economic);
- economic growth;
- development of different types of infrastructure.

Needs of individual members of society or needs of groups that are not identified and communicated within the socio-technical system, will remain unaddressed. For example as long as the real need for reduction of CO_2 emissions is not rec-

---

[56] Glossary.

ognized by organisations and society itself, they do not include this issue in the decision-making process regarding technology development and application.

Even when real needs are identified, there is no interaction possible between needs and resources without positive signals of actors based on social values, interests and visions. Only under certain conditions, the perceived *needs are selected* by individuals, teams and organizations for further processing. For example countries that have identified the need for CO_2 emissions reduction and signed the Kyoto protocol have planned for very different measures for reducing CO_2. Some of these countries emphasise nuclear power, others biomass heating, water energy, biogas, bio-fuels, etc. *Demands* are considered as perceived needs or needs initiated by the market.

Generation of knowledge and skills are key activities within a socio-technical system that start with an initial input and evolve over time. An example is the development of the polymer industry. The knowledge on this new type of material was partly available in the early nineteenth century. However, it was necessary to integrate complementary scientific disciplines such as chemistry, chemical engineering, material sciences, mechanical engineering, physical engineering and certain practical skills to form the new field of polymer engineering in the twentieth century. The integration of each new discipline and new skill was a new challenge and influenced the development of related technologies.

New mechanisms for knowledge generation have evolved. Modelling and simulation technologies have opened up a new view of the world. Future scenarios of environmental and economic world conditions developed by simulation processes, and massive amount of information available regarding natural systems have introduced a new political power for establishing long-term policies for research and development. Engineers are on one hand working on designing such models and on the other hand are influenced by the results of knowledge generated in this way. Future engineers need different kinds of knowledge and skills and at the same time they should be aware of existing limits for each type of knowledge. The quality and the scope of knowledge and skills that are gained in the laboratories, workshops, garages of autodidact people, in the visual world or in interdisciplinary communications at conferences and discussion forums differ from one another substantially.

One of the most important types of knowledge is perhaps the knowledge to understand complex systems.[57]

---

[57] See (Marczyk 2000).

Box 6

Theoretical and practical knowledge generation

Mechanisms of knowledge generation for technical solutions have changed throughout the history. Technical knowledge was a mystery in the Egyptian culture. This culture had undoubtedly a sophisticated level of knowledge regarding building, architecture and transport of heavy stones, long before the times of Archimedes[58] and his laws of mechanics. The Egyptians had however only demonstrated results of their design without clear description of their basic knowledge on technical solutions. It was Archimedes who formulated the knowledge of the law of the lever and distributed the understanding of mechanics in the ancient culture. The ancient Greek culture, two hundred years prior to Archimedes, was more involved in the generation of political and philosophical knowledge. The life and death of Socrates[59] shows the interaction between socio-political development and knowledge generation. Was it Socrates' faith in knowledge that later gave Plato[60], Aristotle[61] and others the idea to establish schools and teach their knowledge that influenced the history of mankind? Aristotle's art of knowledge generation and scientific concepts primarily through the analysis and discussions about the world without any experimental elements dominated the world of science until the Renaissance. The political use of the dogma principle by Aristotle for thousands of years is an example of the interaction between the elite role of science and the policy maker's power.

At the end of the Middle Age era, Galileo was one of the scientists who introduced the generation of knowledge through experimentation against Aristotle's dogma. He opened a new way for scientific and technical knowledge generation through observing the real world, organising repeatable experiments, building hypotheses and generating new knowledge. In the seventeenth century Newtonian physics improved the idea of experimentation. At the same time the communication of knowledge had gained more relevance and importance. It became possible to distribute the knowledge on Newtonian physics more easily due to the society's acceptance of science and the availability of scientific books.

Experimentation has remained for a long time the main element of knowledge generation, especially for technical solutions. However, experimentation has not been the only way to develop technologies. Knowledge has also been generated from philosophy, logics and mathematics without experimentations, in the real world. Modelling became a way that combined experimentation and non-experimental knowledge generation. Physical and logical models condensed the available knowledge for further analysis. The improvement of information technology in the last decades produced a revolution in modelling and simulation as well as a new wave of theories on knowledge generation.

Selection of a technical approach is one of the main phases of the decision-making process at a company at local, national or international levels. Technical choices are not made only according to their functions and economic benefits. Socio-economic, political, cultural, technical and personal factors influence the decisions both before and after the development of technologies through different mechanisms such as regulations, standards, social pressure, etc. Technologi-

---

[58] Archimedes of Syracuse (287 BC – c. 212 BC).
[59] Socrates (470 BC – 399 BC).
[60] Plato (428/427 BC – 348/347 BC).
[61] Aristotle (384 BC – 322 BC).

cal development within a socio-technical system can be described as an evolution process of technology, perceived needs and demands as well as selection mechanisms such as regulation. Schot and Geels (2007, P. 4) discuss the role of a socio-technical regime for selection of technologies. Such a

> "regime binds producers, users and regulators together. The nature of the binding is not based on direct interaction among the actors, but on the participation in the production reproduction of a socio-technical regime. The regime rule-set is embodied in other things, shared engineering search heuristics, ways of defining problems, preferences, expectations, product characteristics, skills, standards and regulatory frameworks. To conclude, a socio-technical regime carries and stores the how to produce, use and regulate specific products and processes."

One decision factor in a process of selecting a technical approach is called vision. Visions are not considered to be directly involved in the fulfilment of needs but they could accelerate or inhibit decisions, make some technologies more attractive with respect to resources, needs or objectives.

Box 7

Visions

Vision is discussed comprehensively in "Visions of Technology" (Dierkes et al. 1996). The Oxford English Dictionary defines visions as "a mental concept of an attractive or fantastic character; a highly imaginative scheme or anticipation." The word is adopted in different disciplines to study the background of individual activities or behaviour in an initial phase of action, to incorporate direction of decision and action and a destination, to be a strategic statement of a business or a direction for a planning, a desired way for embedding of an idea or design in the future. (Dierkes et al. 1996, P. 19-25) It is also a production view, a design view, a theory-building view (Floyd 1989, 1990) in (Dierkes et al. 1996). The word has also been used to show the origin of cooperation issues between engineers and lawyers. It has also been used in the technology development as points of communication between experts and non-experts through which the abstract technological processes can be better conceptualized and brought up for discussion. Vision also serves to envisage various developmental options and anticipated scenarios for discussion and evaluation, to add meaning to the work of researchers and developers, to relate isolated tasks to other projects and to act as a mechanism for coordinating decisions regarding R&D projects and activities.

Understanding of selection mechanisms for technical options and the influence of different technical and non-technical factors are useful not only to the marketing and management staff but also to the engineers involved in the design and manufacture tasks. The conventional tools used in engineering work to formulate decisions for selection of technical options are feasibility analysis, cost, benefit analysis or comparison of mass and energy consumption among different products and processes. The decision-making tools that are based primarily on quantitative data and neglect the influence of qualitative factors based on social values and personal attitudes can produce unfavourable results. The limitation of decision factors to quantitative elements decreases the complexity of analysis while it increases the risk of sub-optimal decisions. Two examples of methods for a more comprehensive analysis and consideration of socio-economic decision factors are impact and technology assessment (IA and TA). IA and TA are techniques for early learning in the socio-technical system and optimisation of mechanisms for selection of technical options. In the last decades of the 20th century technology assessment was institutionalised in the technology policy in order to investigate socio-economic, environmental and cultural impacts of technologies. Other actors of technology development apply these methods at different intensities and scopes in their decision-making process regarding selection of technical options. A description of these methods is presented in Chapter 7 "Methode to examine consequences of technical solutions".

Figure 1 shows the flow diagram for the exchange of information and other resources in a socio-technical system. This concept will be used in this book to analyse determinants of technology development that are relevant to engineering work.

Figure 1: A socio-technical system with feedback loops and actor involvement, D&D&P: Design, Development and Production as three core functions for engineering which include special design, development and production activities such as generation, selection, and consultation.

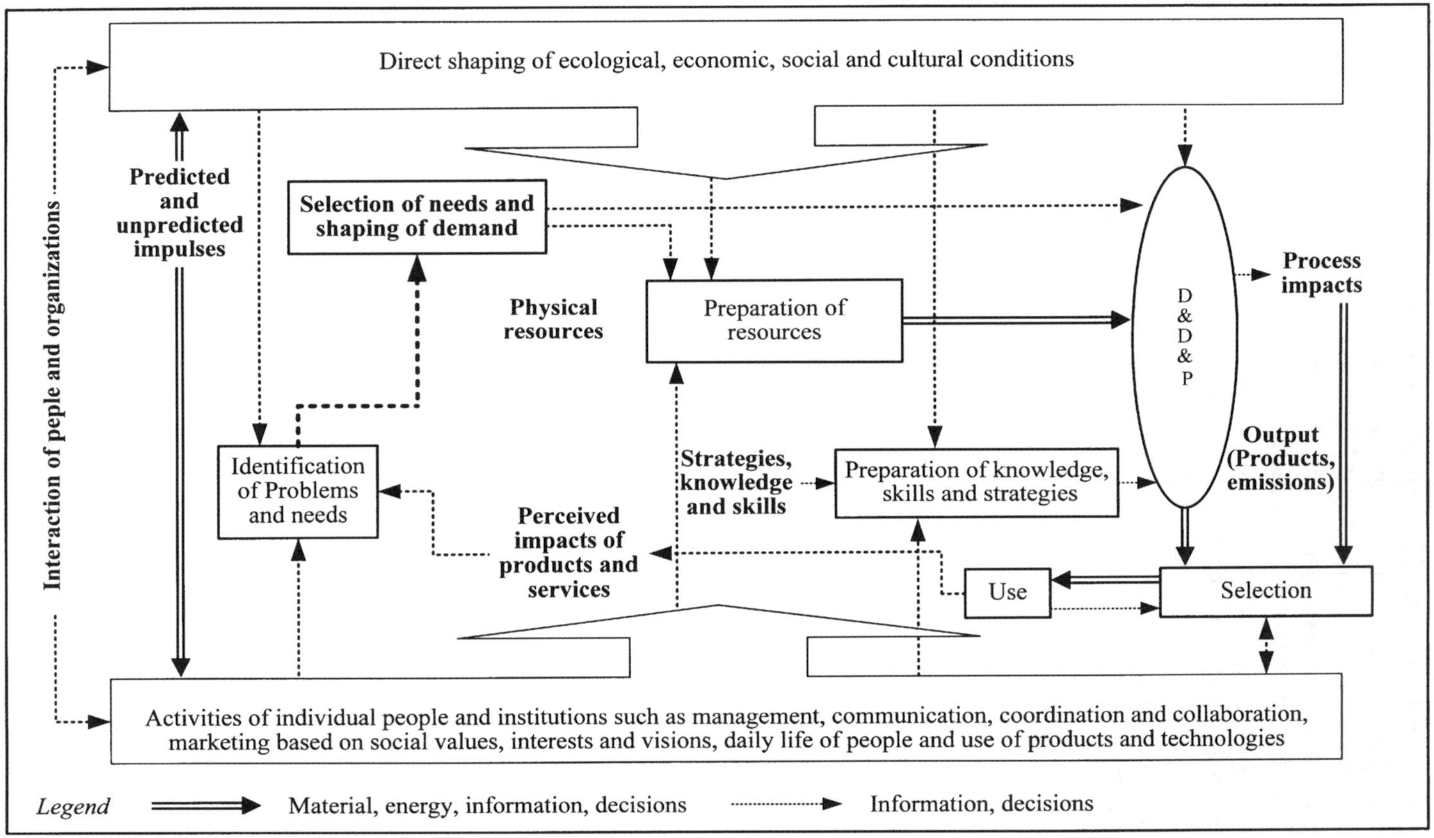

114

Figure 1 shows that needs, problems, resources, rules, interests of actors, networks, impacts of products, processes, etc. are determinants of technology development and application. They constitute the main frame work for engineering activities. Engineers deal with these determinants in different situations, such as:
- established set of determinants;
- changing set of determinants;
- situations with a number of unknown determinants such as unknown impacts of technical processes, products and services.

Knowledge and skills required for engineering work in established situations differ from those capabilities that are necessary for different types of non-established situations or situations with unknown determinants.

The main objective in established situations is the conservation or continued improvement of the situation. This point is viewed as conventional engineering work and will not be discussed further in this book.

In this chapter the author will show some challenges to engineering work due to certain determinants in a socio-technical system. These challenges are combined with uncertainties in regular engineering work but they could also be present in situations with high innovation potential.

According to the concept in the Figure 1 engineers should be able to deal with complexities within the socio-technical systems, influenced by values, interests and objectives, as well as interactions among different actors. Some reasons for the complexity are:
- the interrelated problems and reflective interactions among technical, economic and natural systems (different dynamics of technical, economic and natural systems, uncertainties due to unexpected feedbacks with undesired impacts, …);
- the broad spectrum of involved actors with different interests and different legitimate needs in multifunctional systems.

The complexity of a system increases the uncertainties for engineering work. Actors involved in technology development and application should deal with very different uncertainties related to unpredictable economic conditions and acceptability of their designed products, as well as unknown long-term impacts of such products on natural systems.

A number of different roots of challenges that are relevant to engineering work are listed below and will be discussed in the following sections:
- interrelated problems and needs;
- multiple legitimate needs;
- unintended and unexpected impacts;
- unexpected changes of determinants.

6.1 Interrelated problems and needs

Problems and risks that should be addressed by technical innovations in the 21st century are interrelated issues involving poverty, illiteracy, shortage of clean drinking water, natural catastrophes, etc. and cannot be solved individually alone.

Some of the global and local problems or hazards are classified by Jischa (2005) into three main categories of environmental, social and economic problems. The following list includes problems such as fresh water scarcity that is not only an environmental problem but also a social and economic problem:

Environmental problems
- climate change by green house gases;
- environmental pollution;
- over-fishing of world waters;
- deforesting of primeval forests;
- fresh and drink water scarcity;
- reduction of biodiversity.

Social problems
- poverty;
- lack of education;
- infectious illness;
- war;
- terrorism;
- economic and digital divide;
- natural and environmental catastrophes.

Economic problems on a global level due to lack of appropriate rules for
- world trade;
- international financing architecture;
- prevention of destruction of environment and social power;
- international competition;
- comparable international tax systems;

Economic problems at local level are among other conflicts about
- land use plans;
- big companies versus small and medium sized companies;
- local infrastructure and networks for economic activities;[62]

Most of the issues in the above list are multi-dimensional problems.

---

62 A local community faces needs of its own community and other communities which are in the same network. They may feel responsibility towards some other communities (a club culture of communities) or oppose the needs of other communities outside of their club.

Balabanian (2006) considers environmental problems as a part of new societal problems parallel to other issues such as health problems due to industrial waste and hazards, psychological/emotional problems due to the substitution of machine values for human values, militaristic problems due to Hi-tech militarisation and social problems due to centralisation.

A simple example that shows the multi-dimensional character of such problems can be seen in the engineering work on infrastructure building. The construction of a local infrastructure is connected to environmental, economic and social issues at local, national and international levels. These interactions can be recognised more clearly if we compare the construction of a water supply system in a city during war with a rebuilding project after the war or compare it with construction of a water supply system in a region in a developed country with traditional water supply systems or construction of a water supply system in a new village. In each case there are other interrelated issues that should be considered regarding the engineering work. This implies the challenge and need for a broad understanding for problem analysis for engineering work to select an appropriate solution for each case. A considerable difference between engineering work in sustainable development and conventional engineering work is the emphasis of sustainable development on a variety of contexts of technology development and application. Engineers who are educated for sustainable development need to search for appropriate technical solutions that cannot be found in conventional handbooks written for normal engineering work. They need skills to comprehend interrelations of different problems and use their engineering knowledge and skills in inter- and trans-disciplinary team works.

6.2 Multiple legitimate needs

Diversity of needs or "plurality of legitimate perspectives"[63] poses an enormous challenge to technology development in a democracy. There is a broad spectrum of needs that are not pre-selected and directed by a special group to a homogenate problem definition. Engineering work which encompasses not only economic but also environmental, social and cultural aspects has a broader scope for selection of needs which should be addressed in technical solutions.

---

63 (Funtowicz/Ravetz 2001) in (Decker 2001): "For an example of this plurality of perspectives, we may imagine a group of people gazing at a hillside. One of them "sees" a particular sort of forest, another sees an archaeological site; another one sees a potential suburb, yet another sees a planning problem. Each uses their training to evaluate what they see, in relation to their tasks. Their perceptions are conditioned by a variety of structures, cognitive and institutional, with both explicit and tacit elements. In a policy process, their separate visions may well come into conflict, and some stakeholders may even deny the legitimacy of the commitments and the validity of the perceptions of others. Each perceives his or her own elephant, as it were."

Technical solutions might therefore be directed to special needs while they neglect other identified legitimate ones. This is a fact that should be recognised by engineers knowing that they cannot solve the associated problems all by themselves. The legitimate needs in a society for technical solutions should be recognised within a decision-making process that encompasses environmental, social, economic and cultural needs. The engineer's knowledge is one of the necessary elements for processes such as development of local, national or international sustainable development strategies and specification of needs for technical solutions. The challenge of "multiple legitimate needs" means that engineers' knowledge and skills should be incorporated in a broad manner in decision-making processes. It also implies the need to know and recognize interests of different actors in engineering work.

6.3 Unintended impacts

Engineers who work in new fields are faced with a high degree of uncertainty about the impacts of technical solutions.

The example of automobile production technology shows that uncertainties are not even recognisable at early stages by technology developers. For gasoline and diesel cars there was not enough knowledge about their environmental and social impacts at the beginning of such technologies development. In the early twentieth century the main goal of the auto industry was to provide mobility without being aware of environmental and social effects of the new transport systems. Problems such as smog, rush hours, and stress only appeared after mass production and widespread usage of cars.

During the later phases of the automobiles development when the negative impacts were identified it was difficult to change the automobiles development path. Following the fixed objective of mobility by automobiles, other types of transportation solutions were neglected and city infrastructures evolved with long distances for shopping, school, work, etc. Customers' interest in private cars as a sign of prosperity, adventure or power was intensified by marketing campaigns. The lack of customer-awareness of environmentally friendly mobility alternatives and the established infrastructures are now two main barriers for development of transportation solutions with higher environmental standards.

The problems associated with this type of uncertainties could be reduced by diversity of developed technical solutions, continuous studying of potential impacts of innovative technical developments and the openness to move to better solutions.

Another type of uncertainty for engineering work is an apparent uncertainty that is caused by isolation of engineering work at companies and neglecting the knowledge of other actors beyond the borders of companies. Hints for identification of technical solutions impacts can be found in the works of health and environmental institutes, NGO-reports, customers' associations, etc. These sources should be studied to understand their arguments regarding impacts of products and technical processes. For example, the production and use of toxic chemicals in consumer goods and the pollution caused by industrial processes have been harshly criticized by these actors because of the harmful effects of toxic chemicals on human health and natural system. Medical professor Paul D. Blanc writes in his book "How everyday products make people sick" that reports on the problems of hazards in manufacturing, transportation systems, energy production, etc, dating back to almost 700 years ago. Discussion of these issues is however rarely or poorly present in the learning programs for engineers. Some examples show that harmful impacts are discussed by actors other than engineers after serious dismay of the society. One of such examples is the ban on the use of coal in London in 1316 by King Edward II due to the petition by parliament that saw coal use as a public nuisance. "Edward decreed that those who burned coal should be fined, and on second offence, the burner of coal was to have his furnace demolished, a fairly definitive abatement strategy" (Blanc 2007, P. 14). Blanc gives many examples, which have repeated themselves in the past and modern times, such as the illnesses caused by mercury usage, contamination of freshwater reported in 1868 by Britain's Fisheries Preservation Association, acid rain problems caused by industrial activities reported by Dr. Thomas Scattergood in 1886, pollution of oceans and shorelines by oil spills and the consequent birds death reported by Neil K. Adam in 1936. Blanc concludes that often the old toxins are used again and again in new forms with even more fatal effects. Blanc also provides evidence of the indifference of industry to reports on health problems and fatal cases due to asbestos between 1900 and 1970. His research shows that even today asbestos is used in some countries without any protective regulations.

There are also examples of technical solutions with negative effects when misused or overused. In his work Blanc gives evidence for Carpal tunnel syndrome of postal workers as a result of improper design of machine keyboards in the U.S. post offices for letter-sorting, which required excessive force to operate and were not ergonomically positioned for the workers' operation.

The damage of nerves and body caused by the repeated use of arms, legs, hands and fingers or physiological stress is not a new problem. Dress-makers, carpet layers, and other handcrafters and even Neolithic stone workers have suffered from such painful conditions. A new phenomenon however is that mechanical machines which should help humans and make work easier, lead to the same health concerns today due to mis- or overuse.

The challenge to engineers is in this case to be aware of different needs and problems that are documented in non-engineering sources and implement such ideas in the engineering work. Engineers, therefore, need a thorough education on searching literature in different scientific fields.

The discussion in this section showed that unintended impacts of production processes, services and use of products cause two types of challenges for engineering work. The first challenge is to identify potential unintended and undesired impacts and the second challenge is taking part in the decision-making processes on tradeoffs between benefits and risks of different technical options with different potential impacts. Engineers engaged in decision making and trade –off analysis between economic, environmental, social or cultural risks and benefits have to inform public or private communities about their knowledge on potential impacts of technology and to be ready to discuss such issues with non-engineer experts as well as lay people.

6.4 Challenges due to unexpected changes

Challenges caused by the rapid change of determinants such as resources, needs, or objectives and structure of the socio-technical system are examples that will be discussed in this section. The challenges are of high relevance for the 21st century towards sustainable development.

The challenges discussed in this section are categorised into the following three groups:

I.   changes of inputs such as needs and problems, resources, knowledge, skills;

II. changes of objectives due to changes of internal rules of the socio-technical system or different interests and controversial values, diverse objectives and their interactions;

III. changes of communication mechanisms between people and organisations; formation and changes of networks for new technology development or technology application.

Ad I) Examples of changes of inputs to a socio-technical system are unexpected changes of needs, decrease or change of available resources or change of knowledge and skills. The unexpected changes of needs or appearance of new problems and needs might lead to rapid changes within the established technical systems. The possibility of changing resources could be regarded in some situations as predictable. In other cases such as natural catastrophes, wars or economic crises the loss of conventional resources could shock the socio-technical system leading to massive changes in technology.

An example is the development of rubber technology that was changed by politically motivated resource scarcity during the Second World War.

The example of the rubber campaign illustrates clearly the problem of natural rubber scarcity in the USA during the Second World War. In 1942 the isolation of rubber exporting countries from the USA and the blockage of access to rubber sources due to the war was announced as a serious problem for the country by Franklin D. Roosevelt, President of the United States of America between 1933 and 1945.[64] Rubber was at that time a key product for transportation, insulation of wires for electricity transmission, war armaments, foot wear, adhesives, etc. Because of the scarcity of natural rubber and the high demand for it, synthetic rubber production expanded within a short period of two years, although the properties of the new products were not as good as natural rubber. Synthetic rubber had been known since 1875 in the USA but it was of no interest to the industry due to its inferior mechanical and physical properties compared to natural rubber. After the introduction of different kinds of synthetic rubber to the market at lower prices the natural rubber was used only in special niche markets. Similar examples can be found for energy resources. The two politically motivated oil crises in 1975-1976 and 1979-1983 produced the initial momentum for the development and application of alternative energy resources in industrial countries.

Uncertainties associated with availability of resources have been in many cases the initiator of technical innovations. Resource efficiency and substitution strategies have been two primary measures to reduce the short-term uncertainty. In such cases the products, such as tyres, remain mostly the same. There is evidence that in the cases where resource availability decreases rapidly with time, it leads to fundamental functional changes in the socio-technical system, with phase-out of old products and development of completely new product designs. The scarcity of the "space resource" in big cities has led to construction of tall buildings with new structural principles and materials that are not comparable with single story buildings.

Engineering work has been involved in the examples above in different activities during the change process. Knowledge and skills on conventional technical solutions are too limited to deal with the challenges in such complex situations.

Ad II) Unexpected or rapid changes of social values, objectives and rules in a socio-technical system could make a favourite solution undesirable. For example the toxicity of applied pesticides in the agriculture depends strongly on the environmental consciousness of society. It can be changed rapidly by appearance of scandals on food contamination.

Objectives of a socio-technical system have different scopes. A need fulfilled by a technology is desirable for a specific purpose and a specific group, or for the majority in a very limited field of application. Although established standards keep the direction of objectives and strategies of technology development constant and conserve the development path, differing interests often emerge in the process. In the absence of standards or through an increase in controversial interests it might lead to unexpected changes of objectives for technology development and application. The role of visions can be discussed in this relation.

---

64 Http://www.presidency.ucsb.edu/ws/index.php?pid=16272, accessed 13 May 2008.

Each actor in the decision-making process refers to its own knowledge, interests, values, visions and experiences as well as to objects and other actors (see Dierkes et al. 1996). Dierkes (1996, P. 53) argues that visions are a catalyst for keeping technology development on a special path.[65] However visions are not always clear and transparent. Different visions are often hidden in different stages of the technology design process and can be in conflict with each other. They are one of the sources of unexpected or rapid changes of objectives for technology development and application.

The following example shows the need to evaluate the consequences of objectives for technology development and application. For example, if engineering work is assigned to the primary objective of substitution of fossil fuel with biomass and this objective changes later to become mass production of bio fuels, it could lead to the design and application of technical solutions for mass production of biomass that could result in intensive and mono cultures in the agriculture. The result would not be positive for sustainable development, although the primary objectives aimed to reduce CO_2 production.

A similar example is the challenge to substitution of synthetic polymers with bio-polymers in combination with other objectives. If the substitution of fossil materials is combined with the objectives of awareness-building of consumers for reduction of over-consumption, improvement of efficiency and avoidance of intensive agriculture, the solutions might be positive for reduction of CO_2. If the production of bio-polymers is combined with the secondary objective of cheap mass production, there would be a challenge to engineering work to avoid negative environmental impacts of the corresponding production process.

One of the most volatile conditions that leads to changes of objectives in a socio-technical system is the market condition. A new technology in the market may change an entire generation of products and processes. A very good example is the electronic technology that has changed office work:

Electronic technology both accelerated and slowed down the typewriters development. The introduction of electronic typewriters with a much better quality than mechanical machines was assumed to be a big market success for the typewriter

---

65 Visions promote and constrain the selection of options by practitioners in the overall process of technological development. Visions can be used to make new knowledge conceivable, mobilise producers of technical knowledge and under favourable circumstances they also harmonize interactions between different institutions with different "knowledge cultures". An example can be seen in the development of Diesel engines. Dierkes (P. 83 1996) claims the vision of the Diesel (the inventor) to be the building of the perfect engine after the Carnot theory of heat engines, in contrary to the firms cooperating in the Diesel group between 1887 and 1900 that aimed to build a rapid serial production of stationary engines. The differences in visions slowed down the development of Diesel motor and Diesel group broke up in 1900.

producers at the beginning of 1980s. Developers of typewriters decided based on their long-term experiences with typing machines to invest in electrical typewriters, which were estimated to have a high market potential. By the 1980s, however, personal computers largely took over the market of typewriters due to lower costs, high quality printer technology and the added advantages of calculators.

Engineers who were aware of new market conditions could react to new business requirements and started a new career in designing personal computers. There are also examples of technical education programs that reacted to the innovation needs by integrating product design and planning of multifunctional systems into their theoretical hardware and software courses.

Awareness of the consequences of changing the objectives for technology development and application is a private and personal issue. According to the industrial rules engineers who are engaged in industrial production are expected to estimate potential rewards and risks of different technical options. According to the precautionary principle they have to estimate the consequences of their technical solutions. Engineers however have never been expected to estimate consequences of changing objectives for technology development or application. This is indeed a task that implies inter- and trans-disciplinary work that should be performed with participation of engineers, stakeholders such as companies, politicians and citizens who are affected by decisions and applications associated with the new technologies. In this case engineers are expected to take part and support such evaluation processes.

Trade-off between economic and non-economic benefits and risks of different technical options:

In some cases objectives for the technology development or application change after unexpected and undesired impacts of technologies are detected. As a priori statement: *it is a mistake to believe that a technical system could be designed and implemented with zero unwanted impacts*.

New technologies and products might have undesired impacts that become apparent later. The social values might change in this case in a phase, where the changes of the products, processes or services are not as easy as they would be at an early stage of the development.

The challenge of engineering work in this case is not only a continuous monitoring of impacts of technical solutions, but also engineers' participation in potential impacts estimation and contribution to decisions regarding further development of technologies. All such activities require a high level of non-basic engineering skills for communication, decision-making and co-operations with non-engineers.

Ad III) The literature survey and empirical examples underline the importance of networks for the research and development process.

The formation of formal and informal networks and cooperation[66] at various levels is an organisational challenge that is necessary for the integration of technical innovation:

> "Technical innovations always originate in a given network. Each network features a different constellation of social actors, including universities and public research institutes, various departments in the plants of the manufactures and commercial users, associations, bodies responsible for international standardization, and customer organizations and each involves divergent, often conflicting sets of interests. All these factors largely determine which fields of possible developmental paths are seen as a new technology emerges, which of these paths are perused and which alternatives are ignored or abandoned."

(Dierkes et al. 1996)

There are many examples showing that a critical step in technology development is building interfaces between activities of different actors. Each organization has its own specific rules and structure that are often incompatible with other organisations. Small companies, international enterprises, individual consultants, academic institutes, research institutes, local communities, federal ministries, non-governmental organizations, etc. all have their own missions, rules and cultures. As long as they do not take part in a communication processes with other parties they can use their internal rules. Such entities will have to deal with the challenge of following external rules as soon as they start cooperating with others. On the issue of network building, we should start with the statement that there is no clear identification for the factors that generally cause or induce the change in or development of networks.

The origin and development of networks differ from case to case and the pertinent patterns seem to be "idiosyncratic". There may be a wide and stable network of people, as well as a short-time connection between some researchers for a concrete and limited research project. It is important to know the variety of network types, because it indicates that there are no theoretically-based patterns that would predict a "successful" innovation-network for a new technology. A brief review of issues "networks", "clusters" and "platforms" are presented in the Appendix C.

Engineers often work at different organisations with different decision-making mechanisms and cultures. Today networks with multiple participating groups working together at international, national, regional and local levels are important elements of technology development and engineering work in order to build more flexibility beyond the borders of each individual company. Networks support the interfacing function and at the same time they add an explicit cultural dimension to the technical innovation.

---

66 A brief description of some relevant networks for socio-technical systems and cooperation are addressed in the Appendix C under the topic "Networks, clusters and cooperation".

In summary, changes in a socio-technical system such as changes of inputs, social values and objectives, diversity of structure of organizations and their cultural environments imply flexibility of engineers, their knowledge on determinants of technology development and application and their ability to cooperate with other actors in a socio-technical system.

In the next section another challenge to engineering work with high relevance to sustainable development is discussed. This is the challenge of engineering work with unknown determinants of technology development and application.

6.5 Trade-off for decision-making in a socio-technical system

The common characteristics of all discussed challenges so far are the need for trade offs in the engineering work.

The trade-off is necessary at different levels:

- at company level for comparison of some products or technical processes;
- at local level for selection of a best solution;
- at national and international level for decision on a certain technological development.

Different intensity of uncertainties is possible at each of these levels.

At company and local levels with low uncertainties, it is possible to perform tradeoffs with standard methods such as Life Cycle Assessment and Environmental impact assessment. These methods are however not applicable at higher levels of decision-making involving more complex systems and at higher level of uncertainties. These situations are discussed briefly in the next two sections:

6.5.1 Examples for challenge of trade-offs with low uncertainties

There are two broadly accepted decision-making tools for product development and projects execution. The Life Cycle Assessment for a products development and the Environmental Impact Assessment for complex products and infrastructure development projects such as design and construction of recycling and waste-water treatment plants or airports. Engineers play different roles in such projects. They act as managers of the research and development departments, designers and managers in small and mid-size enterprises of producers or suppliers, consultants or administrative staff.[67] In all these positions they can be involved in such assessment processes. The concept of Life-Cycle Analysis (LCA)

---

67 In VDI Wissensforum (2005) on favourite careers of engineers in Germany more than 50% of participants named a career in research and development as their favourite choice.

of products and processes[68] includes the consideration of material and energy consumption as well as certain impacts (such as loss of biodiversity, climate change, etc.) for raw material extraction, preparation and conversion, production processes of different elements in different industrial sectors, transport of elements to the integration process, use of products as well as repairmen and recycling steps.

The concept of Environmental Impact Assessment (EIA)[69] not only takes the LCA aspects of a project into account but also involves a comprehensive evaluation of industrial products, processes and services within diverse projects at local level. Evaluation of industrial projects often requires decisions on acceptable levels of emissions and risks. Acceptability is however an issue outside the realm of engineering expertise. Even the integration of additional knowledge, such as biological mechanisms of natural systems or social and cultural mechanisms with engineering knowledge is not sufficient for making decisions regarding the acceptability of risks. The engineers' knowledge provide in such cases only input to the discussions and negotiations between relevant actors. Engineers need negotiation skills for communicating with their local community in such cases.

6.5.2 Trade-off with higher uncertainties

The performance of a trade-off method in the decision-making process depends on the quality of the initial estimation and evaluation of potential impacts. One concept for dealing with uncertainties in estimating the potential impacts is the concept of precautionary principle. Precautionary principle in environmental policy has pragmatic and ethical dimensions. The ethical dimension is based on assuming responsibility for future generations. The pragmatic aspect contributes to the definition of rules and standards based on the newest natural sciences knowledge on long-term environmental protection and political decisions for long-term environmental policy (see Kohout 1995).

Precautionary principle in sustainable development is defined as Principle 15 in the Rio declaration on Environment and Development 1992[70] as an approach to start cost-effective measures in order to prevent environmental degradation in the presence of risks with serious or irreversible nature even in the absence of full scientific certainty.

Precautionary principle is interpreted in different ways. It can be regarded as a strategy to use cost-effective measures to slow down or prevent developments with high potential risks, while it can also be used as a framework for continuing the development of certain innovative technical designs even without full scien-

---

68 LCA is described in chapter 7.
69 EIA is described in chapter 7.
70 (UNEP 1992).

tific certainty. In the discussions on controversial issues, when experts' opinions regarding the potential impacts of technologies development and application differ strongly the differences of opinions become more explicit and clear. Experts' findings and opposite opinions may lead to controversial discussions and conflicts. For example, in the case of gene technology development experts such as scientists, engineers, technicians, industry representatives, consumer groups, administration, and staff etc. shape and intensively influence the controversial debates on the technology. Some of these experts view gene technology as a cost-effective measure with environmental benefits that should be followed even with the lack of full scientific certainty. Other groups regard the technology itself as a serious or irreversible threat. Development of precautionary policies on controversial issues and making decision regarding measures in a socio-technical system depends strongly on the quality of discussions and contrast of pro- and contra arguments.

Thomas von Shell claims the central questions on controversial discussion to be:
• Which solutions are available for such controversial issues?
• Is it possible to change positions in the controversial discussion?
• Who are the actors involved in debates on technology and its relevant controversial issues? (Schell 2001)

Discussions and negotiations have a key role in the decision-making process of controversial issues. Making decisions about the application of technology needs a consensus in order to define accepted risks and deal with responsibilities for potential impacts. Thomas von Shell explains the contrast of non-experts real world experiences against expert knowledge and experiences as a positive support for policy and decision making processes. Experts answer the questions "What is possible?", and "What is feasible by the new technology?" The society defines "What is desirable?"

Precautionary principles are however viewed differently in the declaration of a scientific community in the Lowell statement[71]. Scientists attending the "Lowell International Summit on Science and the Precautionary Principle" in September 2001 declared their opinions about ecological modernisation and used the precautionary principle as an innovation determinant. At the same time the statement presents a clear scepticism about the public's different understanding of these issues. The statement presents a fairly idealistic perspective and defines the effect of uncertainties as an initiator for the development of innovative technical solutions in a sustainable development:

> "We understand that human activities cannot be risk-free. However, we contend that society has not realized the full potential of science and policy to prevent damage to ecosystems and health while ensuring progress towards a

---

[71] Glossary.

healthier and economically sustainable future. The goal of precaution is to prevent harm, not to prevent progress. We believe that applying precautionary policies can foster innovation in better materials, safer products, and alternative production processes."[72]

The Lowell statement describes risks and uncertainties as an initiator of change through applying precautionary policies but does not give a clear answer to the question about the definition of an appropriate decision-making mechanism. It does not focus on controversial discussions and the role of society in the decision-making process.

The variety of interpretations of precautionary principle makes it necessary to use a transparent and clear strategy for decision-making under high uncertainty and potential risks.

Based on the discussions above, the decisions regarding precautionary policies should be made based on up-to-date and comprehensive analysis of the related multi-dimensional impacts. And it should be performed in a framework of transparent and potentially controversial discussions about uncertainties.

The decision-making processes for precautionary policies could be supported by a real time[73] impact assessment during technology development and application. A real time impact assessment purpose is to provide up-dated analysis and identify potential weak points of the technology at the early stages of development.

The example of automobiles can be used here to show the need for early warning. The main goal of the auto industry in the 1930s was to improve the transportation speed and to benefit from the technical and economic potentials of new technologies and organisational structures for innovative vehicles for all people. The safety and environmental impacts of automobiles were neither discussed nor identified. Huge initial investments in the infrastructure and design of automobiles have made making corrections and structural changes for mobility with less harm to the environment difficult.

The question of whether an impact assessment should be performed by engineers or not is interrelated to the question of responsibilities of engineers during the life cycle of technologies as well as the question on cooperation and communication among different actors in a socio-technical system.

6.6 Some conclusions for engineering work

This chapter presented a concept for a socio-technical system for the discussion of challenges to engineering work in the 21st century. The chapter also showed a

---

72   Http://www.biotech-info.net//final_statement.html.
73   See Section 7.4.1.

basis for the identification of relevant elements for discussions about technical universities in chapters 9 and 10. The concept is developed based on understanding of technology development as a social process. Analysis of all elements and functions associated with the presented concept are beyond the scope of this book. The focus in this chapter was on the challenges to engineering work based on changes of determinants of technology development and application as well as challenges due to uncertainties. Diverse activities in a socio-technical system imply new skills beyond the basic engineering skills (for example trade-offs between benefits and risks of different technical options).

The controversy in pro and contra arguments on risks and benefits of the technology is mostly a conflict of disciplines. The pro and contra arguments on impacts of climate change since the 1990s among experts (including engineers) show that most of the controversial opinions are present at the boundaries of disciplines and organisations. Taking part in decision-making processes and discussions regarding controversial issues in participatory processes implies new challenges to engineers and their needs for gaining new skills. Technical universities are a key actor in the training of engineers and technicians and teaching learning of these skills. They are important institutions for creation of knowledge and shaping engineers' skills on inter- and trans-disciplinary work and gaining experiences with trade-off between the multi-dimensional risks and benefits of technical options.

7. Methods to examine consequences of technical solutions

The discussion of the socio-technical concept in chapter 6 addressed different activities and determinants of technology development such as potential impacts of technical solutions on society and humanity. The examination of consequences of technical solutions is a crucial research and policy field today.

In the past the manufacturing and technical processes with negative impacts were banned after the authorities' or the public's extreme dissatisfaction. Boiler-construction practices were evaluated in the USA in the early nineteenth century and regulated circa 1850 after a series of steamboats boiler-related accidents.[74]

In 1910 asphalt production was banned in Hötting/Austria to protect the air quality at a popular tourist location (Noggler 2001).

The environmental crises in the 1960s and the expansion of public-funded research led to the development of an institutionalised quality assessment of technologies for early warning. Research fields such as Environmental Impact Assessment (EIA) and Technology Assessment (TA) emerged in the 1960s-1970s and tools such as Product Life Cycle Analysis have been used in the last decades to support decision making on selection of products and production technologies.

Today there are a large number of national and international guidelines and tools for quality control of technical processes, which are used for small-scope decisions at companies or large-scale decisions regarding public infrastructure systems. Figure 2 gives an overview of such issues and their interrelations that will be briefly discussed in this chapter.

Figure 2: Impact assessment for R&D policy

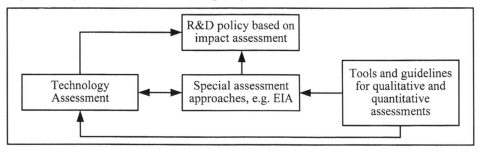

---

74 "In the early 19-century Steamboats frequently blew up when the boilers weakened with age or were pushed too hard. After several particularly murderous accidents in 1816, the city of Philadelphia consulted with experts on how to design safer boilers, the first time an American governmental institution interested itself in the problem. ... It took from that first inquiry in 1816 to 1852 for Congress to pass effective laws regulating the construction of boilers. In that time 5000 people were killed in accidents on steamboats." (Feenberg 2004).

Impact assessment is also a part of Quality Management Concepts that have been developed and applied by the private sector. A comprehensive approach is the Total Quality Management (TQM) that is geared towards achieving quality of products, quality of life at work and outside as well as quality of shareholders' satisfaction. The TQM concept was developed originally in the USA but was implemented primarily in Japan until the 1980s. The Global Environmental Management Initiative (GEMI) has developed a special approach for Total Quality Environmental Management (TQEM) that adds risk assessment and management, environmental auditing programs, environmental sound design, voluntary environmental labelling programs, etc. to the conventional TQM concepts (see Shen 1999). International standards ISO 14000 to 14025 refer to the elements of TQEM. Learning and understanding of these approaches can be particularly achieved today by taking distance learning courses, for example.

This chapter will focus on Impact assessment approaches which are used both for public and private applications. The next section includes a brief history of Life Cycle Assessment as one of the first tools with a high distribution grade in decision-making processes.

The sections thereafter will address more comprehensive methods such as Technology Assessment and Environmental Impact Assessment as well as a new method that is being developed called Sustainability Assessment of Technologies.

7.1 LCA a Widespread Tool

One of the basic and widespread methods for the design of environmentally compatible products is the Product Life-Cycle Assessment (ISO 14040, 14041, 14042, 14043) that is briefly presented in this section. This is mainly a quantitative approach that is developed by natural and engineering sciences.

Beginning in 1970s there have been efforts to develop Product Life-Cycle-Assessments methodologies. The first generation of standard methods was developed by the Society for Environmental Toxicology and Chemistry (SETAC) in 1993. It was introduced as an objective process under the key word product "Life Cycle Assessment" (LCA) to evaluate the environmental burdens associated with a product, process or activity by identifying and quantifying energy and materials used and wastes released to the environment. LCA was one of the first tools for selection of products and processes in the industry. LCA is composed of four components:
• Goal definition and scope
• Inventory analysis
• Impact analysis
• Improvement analysis

Jackson (1993) describes the method as a holistic environmental accounting procedure which quantifies and evaluates all wastes discharged to the environment and energy and raw materials consumed throughout the entire Life-Cycle, beginning with the mining of raw materials to the manufacturing and distribution for consumer use and disposal.

"It is a method to
- calculate the energy consumption and the environmental releases caused by products (the inventory stage);
- evaluate the environmental impacts of releases;
- seek societal valuation of impacts." (Jackson 1993)

The social valuation is often limited to the valuation of companies working environments.

Shen (1999) described LCA as a tool for system analysis that incorporates raw and finished materials, final and intermediate products as well as ultimate disposal processes and services: "The life-cycle concept is simply the holistic approach to evaluating the environmental impacts of a product system from cradle to grave." LCA implies however availability of a tremendous amount of data. In the Streamlined Life-Cycle Assessment[75] report which was prepared by SETAC North America, Todd (1999) addressed practical problems associated with the first concepts of LCA and among others made suggestions for higher compatibility with the International Standards Organization (ISO) document "Environmental Management-Life-Cycle Assessment-Principles and Framework" which was published in October 1998.[76]

"In 2000, UNEP and SETAC established a joint effort to cooperate for the enhancement of the application of LCA and life-cycle thinking, or, in other words, to bring life-cycle thinking and LCA into practice."[77]

The next stage of development was a design of different LCA-software[78] for diverse applications.

Siegenthaler (2005a) reports that 10% of all companies listed on the stock market in Japan run LCA activities, however discussions on the application fields and limits of the LCA-method are still continuing.

---

75 Identification of elements of an LCA that can be omitted where surrogate or generic data can be used without significantly affecting the accuracy of the results.
76 The Glossary of this book includes terms such as Environment, Life-Cycle Assessment, , Functional Unit, Life-Cycle Impact Assessment (LCIA), Life-Cycle Thinking, Screening LCA, Streamlined-LCA, which are cited from this document.
77 http://www.scientificjournals.com/sj/lca/Pdf/aId/5206, accessed 13May 2008.
78 Such as NIRE-LCA Software developed by the Japanese National Research Center for LCA http://www.lcainfo.ch/DF/DF25/Conference%20Reports.pdf, accessed 13 May 2008 (Siegenthaler/Margni 2005).

Another approach that is more comprehensive than Product Life Cycle Assessment and is applied in sophisticated programs as well as comprehensive plans focuses on the Environmental reporting of activities. This is the Environmental Impact Assessment approach (EIA).

7.2 Environmental Impact Assessment

EIA was initiated and developed in the USA according to the U.S. National Environmental Policy Act, NEPA 1969, which implied the preparation of Environmental Impact Statements for the proposed major Federal actions.[79]

The 1985 EIA Directive on Environmental Impact Assessment of the effects of projects on the environment in the European Union was amended in 1997 and then again in 2003 with regard to public participation and access to justice.[80] EIA can be undertaken for individual projects such as dams, highways, airports or factories construction.

Porter (1980) defined the central questions of EIA as:
- What are the beneficial and detrimental impacts on the physical environment?
- Which adverse effects cannot be avoided and why?
- What alternatives exist?
- What irreversible and irretrievable resources are committed?

Environmental assessment for plans, programs and policies that have not been defined yet as concrete projects are addressed by "Strategic Environmental Assessment" (SEA). SEA-Directive of 2001 is also enforced in the European Union and includes a mix of mandatory and discretionary procedures for assessing environmental impacts.[81]

---

[79] See (Porter et al. 1980): "NEPA was designed to insure that information about potential environmental impacts of proposed projects was available to the general public. This marked the beginning of public participation in environmental decision making and provided the impetus for environmental litigation. Ideally, NEPA is intended not only to require evaluation of potential projects but also to encourage critical evaluation of the environmental impacts of possible alternatives. Actual implementation of NEPA has resulted in numerous court cases concerning procedural as well as substantive questions" (Moyle 2004).

[80] "The objective of this Directive is to contribute to the implementation of the obligations arising under the Århus Convention" (European Commission 2003). The directive covers fields such as hazardous waste, air pollution, package, etc. and all activities which may have a significant effect on the environment.

[81] "SEA will contribute to more transparent planning by involving the public and by integrating environmental considerations. This will help to achieve the goal of sustainable development" (European Commission 2001).

Porter (1980) suggest that EIA can use Technology Assessment (TA) to:

- Describe the technology state of the art,
- Study alternatives to achieve the object,
- Study impacts and consequences,
- Identify interested parties and those affected as well as
- Identify public policy measures to minimize or prevent undesirable impacts.

He suggests combining TA and EIA together to have a systemic approach of assessment with emphasis on the role of interests and identification of affected communities.

A new concept which is being developed for policy consulting is called "Sustainable Assessment of Technology" (SAT). The approach is described briefly in the next section.

7.3 Sustainable Assessment of Technology

In the last decade intensive research has been performed on environmental, economic, institutional, social and cultural indicators for evaluation of activities according to their influence on progress towards Sustainable Development. Indicators (SDIs) are mostly developed for macro analysis with the need for consolidated data at national level (for physical indicators see glossary).

Pinter (2005) prepared a report on the emerging trends in the field of Sustainable Development Indicators (SDIs) for the United Nations Division for Sustainable Development (UNDSD). The report points out the institutional, methodological and technical challenges of sustainability indicators that measure the progress at sub-national, national and international levels towards sustainable development.

- The key challenge from an institutionalised perspective is the integrating of SDIs in the decision making processes such as development of government budgets or long-term funding plans.
- The technical challenge is related to the problems with data availability and quality.
- The methodological challenge is caused by the lack of standards and comparability of indicators.

The same problems exist for indicator systems to assess the impacts of technical solutions on the progress towards sustainable development.

The aim of the new concept of Sustainable Assessment of Technologies (SAT) is the provision of standard approaches for integration of environmental, economic and social considerations and considering resource efficiency and social acceptability, not only for short- but also for long-term developments. Technology development is considered, in this case, to be a joint activity within communities and not simply a task for engineers and companies.[82] Public communication and participation is a focal point in this approach. Quantitative analysis is only one part of the tasks. A draft of this approach is presented by Chandak (2006) which shows that engineering design and technology development are completely integrated in a socio-technical system through the implementation of the SAT approach.

A challenge for the assessment remains the combination of different dimensions of the sustainable development concept.

A comprehensive integrative approach which combines different dimensions of the sustainable development concept and could be applied at national level is introduced by Kopfmüller (2001). Kopfmüller uses three basic requirements for sustainable development, namely

- ensuring human existence;
- preserving the potential for production and
- maintaining the development potentials of society.[83]

These objectives are specified by a number of rules. Impacts of an activity can be analysed relative to these rules. In order to manage the complexity, indicators should be defined for each rule. This approach has been applied in different policy analysis studies (e.g. on energy production from renewable resources, cement constructions, water supply systems, alternative motors and fuels, biotechnology in agriculture and electronic waste treatment[84] and organic food production[85]). Kopfmüller's approach integrates setting objectives for technology development with impact assessment. Nevertheless it does not standardise the collection of information and the analysis processes. The Kopfmüller approach describes what should be done but it does not emphasise how it should be done.

Integration of similar approaches such as the method introduced by Kopfmüller (2001) with the SAT could be a new step towards more comprehensive technology assessment and progress towards Sustainable Development.

The SAT and EIA approaches are based strongly on methods for technology assessment. The next section gives a brief overview on Technology Assessment and its importance for engineers (see also Guston/Sarewitz 2002).

---

82 See (Rosen 2002).
83 See (Grunwald 2002) and (Schäfer et al. 2004).
84 (Grunwald 2002).
85 (Schäfer et al. 2004).

7.4 Technology Assessment

Technology assessment (TA) is a research and activity field that covers different types of analyses of technologies impacts on their surroundings. Numerous cases have shown that even useful technologies developed for the improvement of human beings' life standards can have short- or long-term negative and irreversible impacts on human life or the environment (e.g. cooling systems with non-explosive CFC gases[86] were developed to improve safety at home).

The need for early warning of unintended and undesired impacts of new products and technological innovations became more evident during the environmental crisis in the USA. The USA Office of Technology Assessment (1972-1995) was the first institution that received a public budget to conduct an objective and authoritative systematic analysis of complex scientific and technical issues for the Congressional members and committees review.

Technology assessment analysis is applied to:
- identify possible technological options to fulfil certain needs of society;
- explore the environmental, economic, social, cultural and ethical impacts of these options (for conventional and innovative technologies);
- identify uncertainties related to different options;
- contribute to the communication between stakeholders, citizens and politicians to facilitate decision-making on selection of desired technological options.

Most of the institutions engaged in technology assessment are public non-profit institutions (such as Institute of Technology Assessment ITA in Austrian Academy of Sciences, the Danish Board of Technology, Parliamentary Office of Science and Technology in the UK's Parliament or non-profit institutions International Centre of Technology Assessment ICTA in Washington). A number of Technology Assessment institutions in Europe are members of the network of European Parliamentary Technology Assessment (EPTA established in 1990). EPTA members support the parliamentary decision-making process and democratic control of scientific and technological innovations through TA in Europe. Other networks of TA-experts exist in Germany, Austria and Switzerland. The Network of Technology Assessment (NTA established in 2004 in Berlin) covers social sciences on technology research, sustainable development and TA research. Members are individual researchers and practitioners of TA including many engineers.

The interdisciplinary character of TA implies the cooperation of social and natural scientists as well as engineers in TA-studies. Furthermore there are discussions on the demand for integration of some elements of TA into engineering work. In

---

86 Chlorofluorocarbon (CFC) gases were introduced in the early 1930's. Their distribution to the atmosphere was a main reason for stratospheric ozone layer depletion.

the following sections two TA approaches are introduced that are important both for TA practice and engineering work. These are

- constructive TA and a similar new concept real time TA suggested by Guston (2002) and
- participatory TA.

7.4.1 Constructive Technology Assessment and real time TA

One particular form of TA is the "Constructive Technology Assessment" (CTA) which was developed in the 1980s and 1990s in the Netherlands. CTA implies "Early and controlled experimentation, through which unanticipated impacts can be identified and, if needed, ameliorated, and dialogue between innovators and the public, to articulate the demand side of technology development." (P. 98 Guston/Sarewitz 2002). Guston introduces a slightly different kind of CTA, namely the real time TA.

> The real time TA "firstly is embedded in the knowledge creation process itself, it makes use of more reflexive measures such as public opinion polling, focus groups, and scenario development to elicit values and explore alternative potential outcomes. Second it uses content analysis, social judgment research, and survey research to investigate how knowledge, perceptions and values are evolving over time, to enhance communication, and to identify emerging problems. Third it integrates socio-technical mapping and dialogue with retrospective(historical) as well as prospective (scenario) analysis, attempting to situate the innovation of concern in a historical context that will render it more amenable to understanding and, if necessary, to modification."
>
> (Guston/Sarewitz 2002)

One advantage of CTA and real time TA is their focus on the behaviour change of involved people and building awareness of potential unintended impacts and uncertainties during the technology development.

Persuasion of actors for a more critical evaluation of impacts is often regarded as a key success factor to promote environmentally friendly technologies in companies. An example is presented by Vermeulen (1995). He analyses the environmental policy for phasing out the usage of polychlorinated biphenyls (PCBs) during the 1980s in the Netherlands. PCBs were used in ink, lime, rubbers, and pesticides. They were also used as coolants and insulators in electrical equipment. The main policy instrument for the PCB phase out was a subsidy program for companies to help them replace PCB-containing equipment. The direct effect of this policy was evaluated as weak however. The analysis shows that the most important measure for phasing out the PCBs was the persuasion of the target group on the positive value of the environment and their inclusion in the policy making process.

In 1971 an American engineer named Edwin T. Layton wrote the book titled "The Revolt of the Engineers" which presented an engineering philosophy with

138

discussions on ethics and the need for more engineers' responsibility. The mainstream engineering occupation, associations and education institutions have however not strongly integrated reflexive technical studies in their programs.

Real time TA may not be a substitution for other forms of TA, but it is a necessary part of early warning systems and reflexive research for engineering sciences. This research could prepare a valuable pool of knowledge for technical education and could contribute to awareness of future engineers to consider unintended impacts of their designs.

Another function of real time TA is an early warning system as part of the technical research. The history of invention and innovation shows that innovative technologies that are developed for solving an urgent and critical problem were themselves sources of undesired impacts. The example of CFC shows that engineers intended to develop a new system in a short time to reduce risks of cooling systems and to increase indoor-safety of the gas and cooling machines. The risk of ozone depletion and the complex reactions of CFC with UV ray in the stratosphere were not known at that time. A real time TA or CTA could not change the situation, if the actors did not apply the precautionary principle considering uncertainties for absolutely unknown impacts without performing continuous monitoring and research to add new information and knowledge to the available knowledge. The need for precautionary principle surely existed in 1930s for the new and absolutely unknown chemical properties of CFCs and their application and distribution in the atmosphere.

Real time and other types of TA could support building awareness about the important and critical role of scientists and engineers who apply precautionary principle in their work. These people are not well liked in the industry since their criticisms usually slows down the development of products and services. Nevertheless they could accelerate identification of the risks. The use of standard processes in laboratories and quality management of products are good examples of usefulness of such self control processes that have already been accepted.

Scientists' and engineers' critical opinions are essential for TA but they are not the only source of information for TA-studies. Technology assessment today is an interdisciplinary technology research field that covers many fundamental questions on relations between technological and social development in different contexts. Trans-disciplinary approaches are applied in TA to enable the integration of user perspectives and to develop CTA or Innovative Technology Assessment methods.

Participatory technology assessment methods are used in the Netherlands and Denmark to emphasise the learning process of industry and society as one of their main objectives.

In the next section a short description of participatory TA is presented.

7.4.2 Participatory TA (pTA)

Participatory methods had originally been developed to enable society to integrate its requirements into the development process of technologies (Sundermann 1999) and to improve the acceptance of decisions. Dialogue and discussions on different points of views are central elements of these methods.

> "Participatory methods are developed to serve decision makers with knowledge, options, as well as with information for debate and mutual learning. They have been established mainly in Europe, with the aim of 'finding solutions together' or 'generating dialogue' through participatory or interactive processes."
>
> (Nentwich/Bütschi 2000)

Participatory TA approaches such as "consensus conference" and "scenario workshop" were developed in the eighties in Denmark. At the same time the program "Man and Technology–Social Sustainable Technology Design" was developed in Germany to promote critical and rational social discussion on technology development with maximum social inclusion.

Focus group interview is another participatory method, which is often used for analysing the needs and problems of different groups related to certain innovative products and technical systems.

Innovation processes could use participatory methods for other reasons, too. These methods can be used for improvement of coordination of research and development along the supply chain management and entire process or the product life cycle. Although an interdisciplinary cooperation between many groups of experts is necessary for such developments, there are many organisational obstacles to such cooperation (Sotoudeh et al. 2000). Participatory methods could improve co-operative behaviour of actors and interdisciplinary cooperation at an early stage of technology development.

Some examples of participatory methods dealing with innovations are analysed in (Joss/Bellucci 2002). The case studies are classified in "Expert-stakeholder pTA" and "Public pTA". In an "Expert-stakeholder PTA" experts or stakeholders become actively involved within TA-process, while citizens play a central role in a "public PTA".

A practical example of a focus group (a pTA method) that deals with environmental innovations is presented in Rohracher (2000). The aim was to investigate the impact of the design and the chosen energy systems on residents of houses with very low energy consumption.

An example of an international pTA-project is the "Meeting of Minds: European Citizens Dialogue on Brain Science". A citizens' panel consisting of 126 persons from 9 European countries was organised by the new method "European Citizens Deliberations".

140

Technology assessment institutes also use pTA to contribute to public dialogue on different issues. Examples are The Norwegian Board of Technology workshops, open hearings and meetings on different applications of nanotechnology; the Future Search Assessment designed and organized by the Institute of Technology Assessment (ITA) in Austria for public dialogue on energy research program e2050.

7.4.3 The role of engineers in Participatory TA

Some engineering fields such as municipal engineering for infrastructural projects are faced directly with the necessity for public communication.

The practice of municipal engineering incorporates public concerns and values.

"The relationship between municipal engineers and the public is multidimensional, incorporating issues such as:
- multi-stakeholder planning processes;
- demand-side management programs;
- grievance resolution.
Multi-stakeholder planning processes can be invaluable in incorporating public values into complex and difficult issues or decisions facing a municipality. Involving the public in such processes not only leads to smarter solutions, but also helps win public approval for whatever solution is finally adopted. In some municipalities, public processes are used on a widespread basis, particularly when potentially controversial land use changes are at issue."

(Long/Failing 2002)

Generally, technicians, managers and engineers who develop and implement or manage technical systems need inter- and trans-disciplinary cooperation. They are expected to understand non-engineer users' needs and integrate these requirements in the technical design. Participatory TA methods could facilitate the user integration process, only if engineers take part actively in the communication process and contribute to the information exchange.

One of the engineers' roles is to learn about cultural values and ethical perspectives, when they work on socially sensitive issues. They are expected to consider cultural values that are identified through participatory approaches. Participative approaches prepare an arena for debates on ethical issues, different social values and interests, and they contribute towards transparent and open decision-making processes to non-experts. Nevertheless there is usually no guarantee of their contributing to the results of decision-making without commitment from policy makers and engineers.

Engineers are therefore involved in all phases of participatory TA with different roles not only to learn but also to contribute to the knowledge pool and act as promoters and facilitators for consideration of results of PTA in the decision-making process.

7.5 Conclusion

Impact assessment of technologies is a key task in a socio-technical system. It integrates public and private sectors and is institutionalised for policy consulting.

In the last decades a number of tools such as Life Cycle Assessment, environmental quality control standards such as ISO 14000 and comprehensive assessment methods such as EIA have been developed to analyse mainly the environmental impacts of different options and support the public and private decision-makers for selection of technical options.

Technology Assessment is a core for comprehensive assessment methods and introduces the inter- and trans-disciplinary elements to the assessment approaches.

The brief description of TA in this chapter included indications for different roles of engineers as a stakeholder group in TA:
- to support understanding of interactions between technology and society;
- to identify (potential) problems caused by technologies;
- to participate in shaping of technologies based on knowledge gained by technology assessment.

Technical education should therefore improve future engineers' capabilities for these inter- and trans-disciplinary tasks. Another option is to integrate TA from its theoretical and practical perspective into technical education.

Integration of TA into technical universities has however faced barriers. One reason might be that the engineering profession and technical education have usually been a black box[87] isolated from technology assessment. Another reason is the lack of discussions inside the engineering community on shared responsibility for impacts of engineering work beyond the borders of companies.

The most important barrier however is the lack of communication between society and the engineering community on the benefits and risks of technologies and engineering work. This communication with engineers should be a more comprehensive debate than usual discussions with industry representatives and industrial managers. This debate needs to be primarily an initiating procedure.

A broad and deep debate with a substantial purpose cannot take place by individual actors. After a discussion on objectives and measures for technical universities, in the last chapter of this book, we will return to analyse the preconditions of this debate.

---

[87] Glossary.

Part III

Technical education – Technology – Engineering

This book started with the identification of questions regarding the social relevance of future engineering education and continued with key discussions and statements addressing the needs for technical education in the 21st century that are summarised in this section. Other relevant processes such as cultural development are not addressed in this book. Figure 3 shows the relation between the three parts of the book and the central role of identification of different responsibilities for the engineering profession.

Figure 3: Three parts of the book contribute to the Identification of responsibilities for engineering profession.

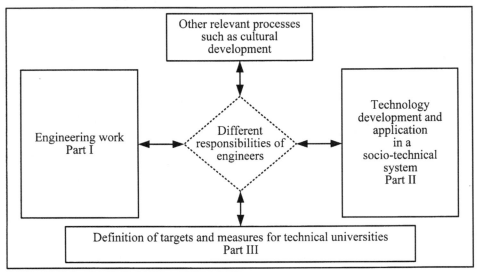

8. The responsibility of technical universities as sustainable universities

To start discussions on technical universities in sustainable development we need to have an overview of discussions on social values for technical universities.

The first declaration on universities in sustainable development, the Talloires Declaration, was signed in 1990. More than 300 college leaders committed to the pursuit of a sustainable future. Other declarations such as the Copernicus Charta followed this movement. A number of universities started developing environmental management strategies for their campuses to reduce energy and material consumption, as well as achieving waste- and emission-reduction. Some universities integrated new courses on environmental management in curricula or established sustainability centres. The terms such as "sustainable universities" and "sustainable higher education" have increasingly been used in different declarations. The next two sections give a short overview on a number of relevant documents for higher education in sustainable development.

8.1 A list of documents for higher education in sustainable development

8.1.1 Talloires declaration (1990)

Talloires declaration (1990) was introduced in 1990 at an international conference in Talloires, France. It was the first official statement made by a university administrator on commitment to environmental sustainability in higher education."[88]

8.1.2 COPERNICUS (1993)

CO-operation Program in Europe for Research on Nature and Industry through Coordinated University Studies was launched in 1988 by the Association of the European universities (CRE). The University Charter for Sustainable Development of COPERNICUS (1993) aimed at fostering environmentally aware attitudes, skills and behaviour patterns, as well as a sense of ethical responsibility.[89]

---

[88] See Glossary.
[89] (COPERNICUS-CAMPUS 2007).

8.1.3 Bologna Declaration (1999)

This declaration by the European Ministers of Education in Bologna fosters the international cooperation in educational fields.[90] The objectives of the declaration are cited in the glossary.

8.1.4 GHESP (2001)[91]

The Lüneburg Declaration of Global Higher Education for Sustainability Partnership GHESP has four founding partners as follows: the International Association of Universities (IAU), the University Leaders for a Sustainable Future (ULSF), Copernicus-Campus and UNESCO – who endorsed the Lüneburg Declaration[92] on Higher Education for Sustainable Development (October 10th 2001). This declaration was addressed to stakeholders and aimed at making education a topic of discussion in the stockholder's dialogue at the United Nations.

8.1.5 UNECE 2005

The strategy of the Education for Sustainable Development of the United Nations Economic Commission for Europe UNECE (2005) emphasises that learning systemic, critical and creative thinking and reflection in both local and global contexts are requirements for action for sustainable development.[93]

8.1.6 UN, 2005 (DESD)

The Plan of Implementation of the Earth Summit of Johannesburg and the UN-Decade on Education for Sustainable Development 2005-2014 (UN, 2005)[94] states:

> "The overall goal of the DESD is to integrate the principles, values, and practices of sustainable development into all aspects of education and learning. This educational effort will encourage changes in behaviour that will create a more sustainable future in terms of environmental integrity, economic viability, and a just society for present and future generations.

---

90 See Glossary, Bologna Declaration, ec.europa.eu/education/policies/educ/bologna/bologna.pdf www.reko.ac.at/universitaetspolitik/dokumente/?ID=681.
91 Http://www.unesco.org/iau/sd/sd_ghesp.html, accessed 13 May 2008.
92 Http://web.archive.org/web/20070216161928/http://www.lueneburg-declaration.de/downloads/declaration.htm, accessed 13 May 2008.
93 18th Principle in Rio declaration: http://unece.org/env/documents/2005/cep/ac.13/cep.ac.13.2005.3.rev.1.e.pdf, accessed 13 May 2008.
94 Http://unesdoc.unesco.org/images/0014/001403/140372e.pdf (Annex I), accessed 13 May 2008.

Values

... Understanding your own values, the values of the society you live in, and the values of others around the world is a central part of educating for a sustainable future. Each nation, cultural group, and individual must learn the skills of recognizing their own values and assessing these values in the context of sustainability.

United Nations history carries with it a host of values related to human dignity and rights, equity, and care for the environment. Sustainable development takes these values a step further and extends them between generations. With sustainable development comes valuing biodiversity and conservation along with human diversity, inclusively, and participation. In the economic realm, some embrace sufficiency for all and others equity of economic opportunity. Which values to teach and learn in each ESD program is a matter for discussion. The goal is to create a locally relevant and culturally appropriate values component to ESD that is informed by the principles and values inherent in sustainable development."[95]

> Joint proposal of Copernicus, GHESP and WWF for the Implementation of the Education for Sustainable Development in the frame of the Bologna-Process (2005).

The suggested principles of action for the transformation of sustainable higher education are listed below:

"1. Institutional commitment
 2. Environmental ethics
 3. Education of university employees
 4. Programmes in environmental education
 5. Inter-disciplinarity
 6. Dissemination of knowledge
 7. Networking
 8. Partnerships
 9. Continuing education programmes
 10. Technology transfer"[96]

The reader should be aware that there are many other declarations which are not included in this chapter.

The Declaration of Barcelona (2004) which refers explicitly to the engineering education in sustainable development is described in the next section.

---

95 Http://web.archive.org/web/20070124035028/http://unesdoc.unesco.org/images/0014/0014
03/140372e.pdf (Annex I: Page 7).
96 Http://web.archive.org/web/20070207222614/http://www.bologna-bergen2005.no/Docs/03-
Pos_ pap-05/041008_Copernicus-Campus_GHESP_WWF.pdf.

8.2 The Declaration of Barcelona (EESD 2004)

The international community of scientists and engineers who are involved in development of concepts for Engineering Education in Sustainable Development (EESD) and have organized EESD conferences since 2002 published the Barcelona declaration in 2004. This community comes together every two years to discuss new developments and strategies for engineering education. The declaration addresses the following topics:

"• The links between all different levels of the educational system;
• The contents of courses;
• Teaching strategies in the classroom;
• Teaching and learning techniques;
• Research methods;
• Training of trainers;
• Evaluation and assessment techniques;
• The participation of external bodies in developing and evaluating the curriculum;
• Quality control systems."

The Declaration of Barcelona does not specify the definition of sustainable development and leaves different interpretations for the concept of sustainable development, but it is an essential step towards a substantial change of technical education for the 21st century and the education of future engineers. According to this declaration engineers should be able to:

"• Understand how their work interacts with society and the environment, locally and globally, in order to identify potential challenges, risks and impacts;
• Understand the contribution of their work in different cultural, social and political contexts and take those differences into account;
• Work in multidisciplinary teams, in order to adapt current technology to the demands imposed by sustainable lifestyles, resource efficiency, pollution prevention and waste management;
• Apply a holistic and systemic approach to solving problems and the ability to move beyond the tradition of breaking reality down into disconnected parts;
• Participate actively in the discussion and definition of economic, social and technological policies, to help redirect society towards more sustainable development;
• Apply professional knowledge according to deontological principles and universal values and ethics;
• Listen closely to the demands of citizens and other stakeholders."[97]

---

97 Http://www.eesd2006.net/index.php?option=com_content&task=view&id=42&Itemid=96, accessed 16 June 2008.

8.3 The concept of technical universities explicit social obligations

In the last ten years technical universities started specifying the objectives of a sustainable university for technical education. Technical universities play a critical role in technology development and education of engineers who influence the socio-technical system.

> "Technical universities bear a special responsibility in the general societal discourse about sustainable development as technical artifacts have a decisive influence on how the relation between man and nature, the economy and basically everyday life of citizens will evolve."
>
> (Corso et al. 2006)

Sustainable development also implies the challenge for technical universities to become a hybrid of organisations close to technical innovation, local development and global responsibility activities, for example in learning networks.

> "Students do not only need facts and figures. Of course they need to know what real sustainability problems are, but they also need a number of skills, which might be even more important than just being aware of the problem. They need to be capable of understanding the relationship between knowledge and craftsmanship and social and environmental situations and the consequences of these. But they also need to be able to work in teams, think of solutions from different perspectives, take opinions of all stakeholders into account and most of all develop solutions that really matter.
>
> To teach them these skills, next to current teaching methods, at least three ingredients are needed: involvement of students in developing education, genuine interaction of students of different backgrounds (interdisciplinary and/or transdisciplinary teams for example) and student organisations to communicate the importance of SD in an informal way (worldwide)."
>
> (Werk et al. 2006)

Learning and group training in learning networks need special organisation of universities. Engineering education needs more and more interdisciplinary work between teachers from different departments. Tutors are especially important for this team work and integration of different generations into the teaching activities. Tutors are an active actor group at technical universities for communication between the lecturers and students. The survey results presented in chapter 4 showed nevertheless that there are divergent opinions from the respondents regarding the importance of student's teaching activities. These differences are partly based on different social values for teaching and research activities at technical universities.

The most important message of social obligations for engineering work is that technical solutions are not the main driver but a part of measures which should be used to solve complex problems such as poverty, climate change, etc.

Figure 4 illustrates the role of technical universities as an element of socio-technical system for solving complex problems in the context of poverty, global warming, etc. The circular-scheme demonstrates the connection between different dimensions of social and technological development towards a sustainable development. Some of these dimensions such as fairness for future generation are shown explicitly in Figure 4 which points out the interaction between technical innovation and education.

Figure 4 also presents the interrelation of technical universities with their local communities for preparation of infrastructure for universities, etc. The influence of local policy varies in different technical universities. The connections of technical universities to global responsibility are available through participation in learning networks. The example of the development of female engineering (shortly described in chapter 3) shows that such networks indirectly affected a transition in technical education and opening of universities to female students. Local conditions and individual people's engagement played a more direct and stronger role in the transition of engineering. Networks however stabilised such changes.

Figure 4: A scheme which shows technical education as an element of problem-solving in sustainable development; Technical education is according to this scheme in a continuous interaction with social development at a local and global level.

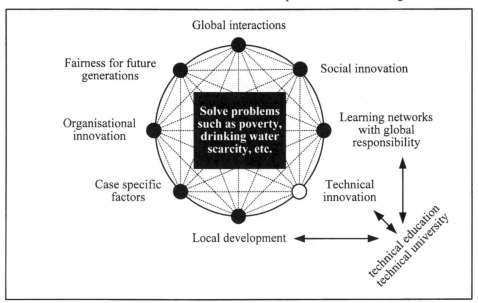

Nevertheless, the conjunctions in Figure 4 between different dimensions should be developed and maintained by relevant actors in each individual socio-technical system. Technical university is in this scheme still a black box. The next section helps to take a closer look at technical universities and specify the cooperation required among technical universities in sustainable development. This scheme will then be modified in chapter 10 to show more interactions between technical education and the socio-technical system.

9. Technical universities needs and objectives

Technical universities are influenced by social values from technology development and engineering profession perspectives. In return, they influence the socio-technical system through generation of knowledge, educating engineers, communication with the local community and global organisations and many indirect interaction channels.

Parts I and II of this book presented the central role of social values in the technology development and engineering work. Figure 1 showed interrelations between social values and technology development that influence selection of resources as well as perceived needs and problems demanding technical solutions and technology development in a socio-technical system. Technology development is shown to be related to:

- engineering work at companies;
- required engineering capabilities and
- technical universities and technical education.

This study shows that:

> "The engineers are expected to be able to take different roles for design, management, development and implementation or service at technical facilities in different countries and regions of the world. They are also expected to be able to deal with communication, coordination, or participation processes for producing and selection of solutions, which go beyond technical factors."

Technical universities are expected to provide adequate knowledge and skills for future engineers and fulfil their organisational responsibilities to society. Due to the case-specificity of factors it is not possible to build a general set of objectives applicable to technical universities. Technical universities objectives that are discussed in this chapter are categorized into four groups as follows:

1. objectives that refer to internal and external integration of universities (refers to organisation);
2. objectives for technical innovation (research);
3. objectives for quality control of education and success of students (refers to education);
4. and last but not least the educational objectives of technical universities that are also part of the other three objectives.

The first and second categories are the starting points for the analysis of technical education and universities needs, in this chapter. Results of the research on the needs and required measures for different changes in technical education and universities will be presented mainly for integration and innovation objectives. The internal quality control and education objectives (the third and fourth categories) will be addressed briefly, since this book presents an outsider view and analysis of such targets require insider insights.[98]

Ad 1) Examples of needs for universities internal and external integrations are:
- integration into the Local community (functional cooperation with neighbourhoods, LA21 and social problems, cooperation with businesses based on the principles of sustainable development);
- interdisciplinary studies and research (survey, Chapter 4);
- environmental Impact Assessment (EIA) and quality control of laboratories activities (Personal communication of colleagues, research results);
- integration of different interests into the decision making processes (principles of sustainable development);
- international cooperation and networks (research results).

After a brief description of these integration issues different mechanisms of communication paths, required in the integration process, will be discussed.

Ad 2) Needs related to technical innovation include:
- better understanding of technical innovation (survey, research results);
- Introduction of engineers to engineering practice early during technical education (Survey, research results);
- Shared responsibility (based on research results should be defined in a communication process with local community including lay people who are affected by impacts of technologies).

The needs listed in these examples have different dimensions. Integration into the local community, for example, has strong economic, environmental, social, cultural and organisational aspects. On the other hand interdisciplinary studies and research have an organisational and scientific character. Environmental Impact Assessment of activities at laboratories has preliminary informational, environmental and scientific dimensions. Integration of different interests into the decision making process has a strong social and regulative aspect. Measures for the multidimensional needs should therefore address multidimensional factors.

The descriptions of measures related to the needs listed above are explorative in nature and are not necessarily a list of the best possible solutions. The text invites individual readers and student groups to analyse the examples, discuss them together and form their own opinions.

---

[98] Educational technology innovation is out of the scope of this book. See e.g. (Abelson/ Long 2008) for remote laboratories.

9.1 Objective: Integration of universities

A main group of objectives address organisational factors. They are underlying elements for education- and research-oriented objectives. The next sections describe some examples for needs and measures which are related to integration of universities objective.

9.1.1 Integration into the local community

Integration of technical universities into the local community has multiple purposes. Some universities are integrated into the city in their neighbourhoods, give services to the local community and share their infrastructure with the city. The universities are integrated into the local community

- physically;
- organisationally by integrating business, or
- deeply through social integration into the city.

Physical integration starts with sharing of resources.

A technical university using energy and material resources strongly influences the socio-economic condition of geographic areas far beyond its neighbourhood in the city. There are a number of activities that can be shared between a university and the city to use synergy effects.

The university can be integrated physically into the city. Some streets can be shared between the university and the city.

Organisational integration is achieved e.g. through sharing university laboratories with new SMEs (Small and Medium Enterprises) with innovative ideas. Small companies would have a "one room office" in the business department, for example, and use the university infrastructure for a certain period of time. This cooperation should be well organised, transparently declared and clearly defined through the university policies and approved by the responsible authorities. The university can also integrate a part of a local technology park and use synergy effects between it and the technology park where students can be involved in hands-on projects and be introduced to SMEs.

The university laboratories can have versatile functions. It is important that they perform based on strict environmental, human and animal-friendly regulations. They should take the necessary actions to prevent toxic emissions, cruel animal tests or tests with human health risks inside or outside of laboratories. Enforcement of such regulations provides selection and permission criteria for innovative small companies that are allowed to use the university infrastructure.

Social integration is a more complex process than physical and organisational integration. A possible approach is the management of the integration process by

153

an office of "LA21". LA21 stands for Local Agenda 21 which is a chapter of a report by the United Nations Conference on Environment & Development in Rio 1992. LA21 builds an important framework for communication at local level and related to local and the international sustainability policies.

> "Because so many problems and solutions being addressed by Agenda 21 have their roots in local activities, the participation and cooperation of local authorities will be a determining factor in fulfilling its objectives. Local authorities construct, operate and maintain economic, social and environmental infrastructure, oversee planning processes, establish local environmental policies and regulations, and assist in implementing national and sub-national environmental policies. As the level of governance closest to the people, they play a vital role in educating, mobilizing and responding to the public to promote sustainable development."[99]

The Institute for Urban & Regional Development (IURD) at UC Berkeley has a similar function:

> "IURD is an organized research unit that brings together faculty and students from across the campus to work on urban and regional development issues. IURD assists the campus community in raising research funds and provides administrative support for this work. IURD also supports graduate education through research training and internships for professional school students. IURD is a primary research home for many UC Berkeley faculties, primarily from the College of Environmental Design, but from other professional teaching programs as well. As a focal point on campus for faculty whose work engages current urban policy and planning issues, IURD brokers relationships with community partners and assists in the dissemination of community based research to wide audiences, in both academic and public venues. IURD has long served as a portal to the university for people in Bay Area communities, including city and county planners, community based organizations and others looking for technical assistance and guidance."[100]

In this section we observed that different measures for physical, organizational and social integration of universities in their local communities lead to different changes at universities. Other forms of social integration will be discusses in the next section. Next section will present interdisciplinary work as an objective that addresses internal social integration at universities and points out the measures required for its success.

9.1.2 Interdisciplinary studies and research (objective/measure)

Interdisciplinary studies and research can be considered as tools for problem solving. They are needed for solving complex local and global socio-economic

---

99 Http://www.archive.org/web/20050507071029/http://www.unep.org/Documents.multilingual/Default.asp?DocumentID=52&ArticleID=76&l=en.

100 Http://web.archive.org/web/20070817150207/http://www-iurd.ced.berkeley.edu/more. htm.

154

and environmental problems such as extreme inequalities in people's quality of life such as poverty or health risks due to food, air and water pollutions. Inter-disciplinary studies prepare engineers for their work and are known (also according to the results of the survey presented in this book) as one of the important success factors for engineers.

Nevertheless, many technical universities still need to improve on interdisciplinary research and studies, since establishment of various kinds of programs for inter- and trans-disciplinary work is a challenge to the traditional disciplinary-organized universities.

A number of universities have added courses for different foreign languages, cultural studies, psychology, moderation and mediation training, international law and regulations for protecting natural resources, etc. to improve students' communication skills for their interdisciplinary works. Sports and art courses are offered to train students on team work and conflict management.

Technical universities may also emphasis an educational program to build a basic knowledge of potential environmental, economic, social and cultural benefits and risks associated with the new and existing products and processes. Engineers with such qualifications can improve their inter- and trans-disciplinary team-work with economic experts, simple workers, technicians, consumers, users of products and other relevant actors.

An important factor for the success of interdisciplinary studies and research is the access to data sources such as books and references that contain interdisciplinary information. University libraries have a key role to promote this measure.

9.1.3 Environmental Impact Assessment and quality control

Experimental work at technical universities can be a major source of negative environmental impacts on their campuses and their surrounding neighbour-hoods. Wastes, emissions, odour or noise reduce the quality of life near such universities. To integrate a university into a city it is necessary to control such undesirable effects and to have a transparent impact activities report by the responsible authorities of the corresponding laboratories within the university.

Some universities have prepared policies that apply to all their research activities. The Research Compliance-Handbook of the UC-Berkely in 2006 is an example of such a policy.

> "This Handbook was intended to be an instructive guide for faculty and campus researchers. It has been designed to inform Berkeley campus faculty members (and other campus researchers) about key "research compliance" regulations mandated by the federal government and by the State of California that require oversight by the campus and additional activities on the part of the researcher.

Included in the Handbook are the following areas:
- Research Using Human Subjects;
- Research Using Animal Subjects;
- Financial Conflict of Interest in Research;
- Research Laboratory Safety;
- Research Misconduct;
- Research and Security Issues in the Post-9/11 Environment."[101]

Environmental Impact Assessment (EIA) is a good example describing the requirements for interdisciplinary research work at universities. In order to achieve a realistic and effective EIA, universities need to promote inter-disciplinary work among all branches of engineering together with legal expertise. There is also the need for trans-disciplinary work to integrate the local community interests through working with the local government authorities.

The University of California at Berkeley example demonstrates the implementation of an explicitly and clearly defined quality management strategy to coordinate and oversee the diverse activities at the university.

9.1.4 International cooperation

Universities need to become internationally active in dealing with major challenges of research on complex social and environmental issues such as health care, environmental protection, mobility, security etc. Objectives such as

"Safeguard the earth's capacity to support life in all its diversity, respect the limits of the planet's natural resources and ensure a high level of protection and improvement of the quality of the environment"

(COUNCIL OF THE EUROPEAN UNION 2006)

can only be achieved by means of international cooperation. Local and national activities need to go hand in hand at international level.

Some technical universities have established separate departments to foster their international contacts and cooperation. The education program in some universities contains training in foreign languages and introductions to different cultures.

International conferences held at a university can be a measure not only for information exchange with the international community but also for training students for their practice at international level.

Students' involvement in such activities is important not only for international cooperation but also for academic performance at universities. Tutors can have an important role in technical universities to improve communication between lec-

---

101 Http://web.archive.org/web/20070816013502/http://rac.berkeley.edu/compliancebook/introduction.html.

turers, research and teaching assistances and students. They can also be involved in decision making processes representing the interests of the student body.

9.1.5 Integration of different interests

Technical universities often apply project-oriented approaches for their research and education programs. Integration of all involved actors interests in decision-making processes can also be conducted in the frame work of special projects. These projects could be selected through implementing comprehensive participatory approaches.

Participative approaches imply, in such cases, cooperation among Student Associations and other representative groups at universities. In some universities students and staff with small children who may stay at home are organised in groups to discuss their ideas in different project teams. Such "reflection teams"[102] are a valuable source for initiation of new research activities.

Another kind of integration of people's interests can be realised through creation of an LA21 office. This office can contribute towards the local community needs assessment and to transfer this information to the university research and education programs. The continuing education program in particular can design student projects based on these needs.

The most important aspect of "integration of interest" is the interrelation among technical universities, industry and public. The preparation and coordination of this kind of social integration will be discussed in more details in the next chapter.

Previous sections demonstrated some examples of internal and external organisational and social integrations applicable to universities. In the next section a discussion about the mechanisms of such integrations is presented.

9.1.6 Mechanisms of interactions inside and outside of the university

Actors' interactions inside and outside of the university can be described by their objectives, paths and contents. Interactions contents are case specific and will not be discussed in this section.

Objectives of interactions

The general objective of the outside interactions is the local integration together with the global engagement of the university in the socio-technical development based on the objectives selected anticipatively.

The communication inside the university is the key element for its self-regulation management with a participative character.

---

102 (Philippi 2005).

Paths of interactions

Outside interactions of a university follow three interrelated paths that can be categorized as organisational, individual and project-oriented. The following definition of paths is based on the results of a study on the implementation of innovative-integrative environmental technologies in 2002 (see Sotoudeh/Mihalyi 2004). These results can be used for development of different kinds of innovative integrative technologies through industry cooperation with technical universities.

Organisational path

The organisational path mainly represents the relation of the university to its related and supporting organisations. Three significant organisational paths for the development and implementation of integrative environmental technologies, which are relevant to technical universities, are:

- Public research institutes could be connected to basic research at the university and applications in the industry through organizational paths. This network can be a very effective way to promote rapid distribution of knowledge and technology.

  "Cassier and Foray emphasize the aspect of the 'collective invention' as opposed to 'privatisation of knowledge'. The collective or pooled data is shared among participants during the period of research."

  (Plunket et al. 2002)

- The second structure represents the connecting role of external engineering firms and technical offices for application-oriented research by universities, manufacturers and users. This variant plays a crucial role for the diffusion of already developed solutions or for process optimisation.

  "Saviotti discusses the importance of innovation networks and the fact that they have become stable components of industrial organisations. He shows that these networks are based on three types of actors, namely large diversified firms, small new technology firms and public research institutions. In particular the stability of these networks is explained by the dual role played by the small new technology firms which both contribute to understanding the technology and communicating it to the large diversified firms."

  (Plunket et al. 2002)

- The third form relates to the new technology firms who cooperate with universities. They can be small companies residing in a university business building or large firms possessing their own process engineering laboratories. A challenge posed by this form of interaction is that the non-disclosure of research results achieved through involvement of innovative firms often causes conflicts related to the diffusion of technology and knowledge.

Individual path

The individual path relates to the people operating within the university organisation and their individual contacts with other organisations. This path is important for the rapid exchange of information and new ideas. To analyse this path it is imperative to study how people behave within an organisation, the reasons for their behaviour, the factors that contribute to their decisions and the methods used for bringing about changes.

Project-oriented path

The project-oriented path sets the physical, administrative, informatics, scientific and technological requirements for information exchange between departments and individuals at the university and its surroundings. Project-oriented path supports both short- and middle term strategies.

> "Successful R&D often depends on linking disciplines and groups with different knowledge and political interests. In a context of high technology R&D, where interdisciplinary projects undertaken by multiple departments, firms and even countries are increasingly the norm, this focus on project work rather than institutions may be the key to successful performance".
>
> (Zabusky/Barely 1997)

The relations within the different paths discussed above can be described as short-, middle-or long-term interactions and they can transform from one form into another. Individual short-term paths can gradually develop into project-oriented paths, which are essential for technical universities. Zabusky and Barley suggest project work as a strategy to successfully promote cooperation among scientists, technicians and non-technical staff from different organizations and disciplines.

The diversity of the interaction paths implies a flexible structure at universities. One of the main results of well functioning interactions is technical innovation by universities conducting research activities focusing on this objective.

9.2 Needs and measures for technical innovations

Technical universities have been a necessary element for innovation in a socio-technical system. Some relevant objectives are described briefly in the next sections.

9.2.1 Better understanding of innovation

Technical innovation needs new ideas, willingness for development and realisation of concepts from ideas and resources leading to usable innovations. It needs not only knowledge and financial resources for experiments, prototyping and tests but also a flexible structure that allows for discussions on different aspects of the new solutions.

Innovative technical solutions can and should be supported by interdisciplinary work in well-designed teams consisting of students, tutors, research assistances and lecturers. Students can learn in such teams various criteria of interdisciplinary works. One such criteria is the critical use of diverse references.

Different understandings of innovation diffusion and evolution which are presented in different references might result in very different plans for action. Technical universities libraries therefore need a wide spectrum of books and journals on understating of innovation diffusion and evolution from perspectives of different disciplines.

There are still many references that use classic economy-based ideas and describe the engineering role only as a utilisation of a physical environment to achieve an economic goal and evaluate the success of innovations only by economic factors. These references do not reflect the recent developments in innovation research. Students should be aware of differences between innovation concept in macro and micro economy and have access to standard books as well as references on alternative concepts of innovation such as the co-evolutionary innovation concept.

Above all, the long-term objective for better understanding of innovation is for technical education to provide knowledge and experiences regarding innovation processes advocating local engagement and global responsibility beyond economical aspects alone (e.g. cultural, social and environmental aspects).

Generation of new ideas and technical innovation are strongly influenced by engineers' practical experiences. Learning by doing will be discussed in the next section.

9.2.2 Learning by doing

Engineers should be able to apply scientific knowledge while maintaining the capability to synthesise new knowledge for their special tasks. They need to be doers and thinkers (see Figure 5). Technical universities curricula need to be rich with projects that combine thinking with doing in engineering. Examples for such trainings can be found in the medical research field where students learn for example through Cardiology Fellowship Research Training that lists

> "hypothesis generation, critical literature review, protocol development, institutional review board submission and purpose, data collection, data analysis, abstract preparation, presentation in scientific forum, preparation of paper for peer review journal submission, working in teams and critically reviewing other researches."[103]

In a version of learning by doing engineering students receive their research assignments that they are asked to work on and solve in teams. They prepare project plans, formulate hypotheses, design experiments, document results and ar-

---

[103] Http://medicine.georgetown.edu/cardiology/research_cfrt.htm, accessed May 13 2008.

rive at answers. They present their results before a jury, prepare a project report and respond to feedback from the jury (see Graaff et al. 2001). Questions for engineering students could be real-life problems related to the local environment or based on global issues with an international dimension. There should also be issues related to visions and long-term strategies. Learning by doing is also a kind of guided self-study that aims at improving the students' problem-solving capabilities. Lecturers sometimes use methods for non-formal learning, as well as creative and collaborative tools in their courses. Social skills for team working can only be improved during the project work and students attain feedback not only for their scientific results but also for enhancing their project management and quality control capabilities. As shown in Figure 5 engineering students can act according to their personal behaviours and external conditions as doers, thinkers, deciders and dreamers at different periods of time.

Figure 5: Kolb's learning cycle in (Graaff 2006)

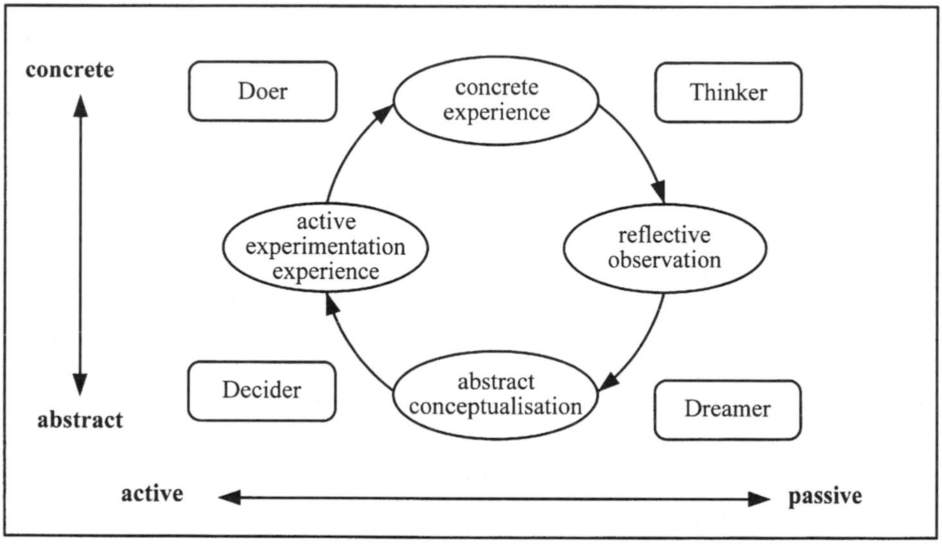

concrete

| Doer | concrete experience | Thinker |

active experimentation experience | reflective observation

| Decider | abstract conceptualisation | Dreamer |

abstract

active ←——————————→ passive

Learning by doing can also be divided into different environments. It can be understood as out-of-campus experiences. "Co-operative education is a rotation program, which alternates between industrial work and academic work" (El-Sayed 2001). These experiences are necessary at local, national and international levels to give engineering students a wider view of serious problems to be solved and the different cultural methods of problem solving.

Learning by doing therefore provides a possibility for engineering students to learn responsibility. In the next section shared responsibility as an objective in relation to the objective for technical innovation is discussed.

161

Engineers need to make decisions in difficult situations and take responsibility for the results of their decisions and the technical systems they develop. Decision-making and management skills should be taught to the broad spectrum of students to make engineers capable of dealing with uncertainties of design and also for taking responsibility individually for the impact of their work. There are also requirements for engineering ethics in the university education program.

Van de Poel (2006, P. 223) gives an overview of the discussion on shared responsibility, which includes external control and regulation, as well as individual responsibility. The practice of shared responsibility should start early at university to prepare students for their involvement in real decision-making processes and projects. Although standard processes and regulations exist at universities that determine the projects staff responsibilities, responsibility still remains individually based on the specific understanding of situations and is a special form of social integration.

It is a part of shared responsibility to understand the interactions between cultural development and technical development. This is required to be able to develop services for civil application with a deep understanding of cultural dimensions at local level while respecting the worldwide needs of human beings. Nevertheless shared responsibility would not be useful, if future engineers do not consider limits of technical innovations, economic growth and social demands with respect to other cultures and societies. Technical education is therefore not only a national or local affair; but also needs to provide international experiences. Engineering students, who spend semesters in other countries with different cultures, have a chance to better learn the different paths of technology development according to different socio-political conditions. Students could learn through these projects the requirements as well as benefits and risks of technology transfer. A part of these projects should analyse critical differences of engineering works in different cultures.

A common objective for universities is their commitment to high quality education and high opportunities for their graduates. Some examples of measures to achieve these objectives are described in the next section.

9.3 Measures for quality control

The quality of technical education depends both on general quality criteria and case specific factors for organisation, education and research at universities. The description and analysis of relevant measures for quality control require mainly case specific and deep knowledge on internal conditions at a university.

9.3.1 Control of scientific and education quality

The quality management of educational and scientific activities at a university has an essential influence on the structure and all activities of the university.

Conventional quality management measures include the evaluation of the course contents, students' learning methods, lecturers' teaching methods and evaluation of their works, departments, etc. by steering committees. These evaluation committees have different levels of involvement of external scientist, politicians and authorities or students and university staff.

Poor methods of quality management that are unfair, inflexible, not innovation friendly or too tough might lead to frustration of the university academic staff and other involved parties. They can cause over- or under challenging activities and might therefore lead to hindering potentials and creativity in generating new ideas at the universities.

The quality management system should therefore be well designed. It should take into account the fact that scientific staff has very different backgrounds requiring individual and specific quality management strategies. There are often a group of engineering practitioners who use a pragmatic and success-oriented strategy in their works, scientists who teach basic disciplinary theories and perform basic research, as well as scientists with a hybrid background who teach the interdisciplinary analyses and research in interdisciplinary teams. The evaluation program has to address the diversity.

9.3.2 Quality control of universities in sustainable development

Sustainable development implies a continuous and comprehensive measurement of success and failure of measures for a continuous quality improvement at universities. In 2001 one of the first assessment concepts was developed in the University of Florida. The quality improvement was defined by environmental, economic, social and education performance.

The selected Performance Indicators that were used in the University of Florida were compliant with the Global Reporting Initiative's publication. These indicators can be used to mark the trends of the performance as positive, negative or neutral:

- Environmental Indicators
  Energy, Material, Water, Transport, Land Use / Biodiversity, Emissions, Effluents, and Waste, Waste Returned to Process (Recycling), Waste to Land, Effluents to Water;
- Economic Indicators
  Revenues, Investments, Wage and Benefits, Community Development;
- Societal Indicators
  Workplace, Health and Safety, Wages and Benefits, Training / Education, Non-discrimination, Freedom of Association;
- Education Indicators
  Faculty, Undergraduate Education, Graduate Education, Campus Safety.

Siemer (2006) has presented the results of a survey which shows sustainability indicator groups used by pioneer universities as follows:

"• Those based on the Triple Bottom Line (TBL) approach;
• Those based on an expanded environmental indicator set (expanded environmental reporting)
• Those based on standardized sustainability assessments, such as the Sustainable Pathways Toolkit (Good Company) and Campus Sustainability Assessment Framework (CSAF) as developed by Lindsay Cole
• Those based on the Global Reporting Initiative (GRI) Guidelines."

Sustainable Pathway Tools focus on 15 indicators on environmental management and health components. CSAF focuses on 175 indicators with ecological dimension with supporting function for the societal sphere.

Universities may use sustainable reporting for internal and external communication. Internal communication through application of indicators contributes to awareness of the university staff and students on the environmental, economic, social and cultural performance of different parts of the university. The external communication through standard reporting systems supports the comparison of the universities performances at national and international levels and learning effect in the universities networks. Even the application of case specific assessment methods has advantages for external communication of a special university with its local community on its objectives and achievements.

The main group of technical universities objectives is educational objectives. They will be discussed as the last group since these objectives constitute a cross-sectional group which has already been addressed in all other objectives.

9.4 Educational objectives

A set of requirements for engineers' education identified in the survey were skills and knowledge for:
• "Communication and presentation skills";
• "Management skills";
• "Basic engineering skills";
• "Cooperation skills";
• "IT skills";
• "Cross-disciplinary work skills";
• "Reflexive knowledge";
• "International experiences and language skills";
• "Practical knowledge gained from learning by doing";

- "Capability of thinking and acting beyond the borders of the company on global and local level";
- "Capability of critical thinking and taking responsibility".

Basic engineering were among the first groups but not the first group of required skills and knowledge. The results of this survey with 44 respondents from different cultures and universities were compared with the proceedings of two different engineering conferences. Both cases indicated the need for inter- and trans-disciplinary work by engineers with non-engineers in teams.

There are already experiences with new curriculum contents for technical education and new programs that are necessary for dealing with complex problems such as climate change, scarcity of fresh water, etc. (see (Mulder 2006), (Bootsma/Driessen 2005), (Jischa 2004))

New methods of learning (e.g. E-learning, inter-cultural dialogue (Dam-Mieras 2006), adult education (Stöglehner et al. 2006), (Corso et al. 2006)) have also been intensively discussed by lecturers and educational experts.

Some measures for changes are incremental and easier to conduct, while others such as comprehensive e-learning programs and learning by doing need radical changes of infrastructure and a well-designed integrative reform plan.

In the concluding remarks some examples will show that management of universities should coordinate implementing these measures considering both the inside and outside factors to assess the quality of performance and opportunities for improvements.

A more detailed discussion on the educational objective is outside the scope of this analysis because this book focuses on an outsider view of technical universities.

9.5 Concluding remarks on needs and measures

The needs and measures introduced in this chapter are examples identified through internet research, interpretation of results of a survey with 44 respondents described in chapter 4, informal personal communications and the author's own experience.

The list can be further expanded by:
- building public awareness on global responsibility for the future of human beings. A scenario can be generated based on well designed lectures for public understanding of science and critical selection of topics on environmental, economic and social issues related to technical solutions;
- finding solutions to individuals' and society's problems to achieve a better quality of life (taking into account chances for next generations);

- emphasising the role of women in engineering with a program that enables women to combine education, career and private life;
- "embedding wide participation and student diversity in organisational practice" of universities as a strategic and corporate policy of the City.[104] The recent research work of the London Academy for Higher Education shows an example for studies to make a change in "diversity of students" of a university. The authors emphasise that a "one size fits all' business case model would not be appropriate. They also show areas for further considerations;
- emphasising health in the quality of technical education and technician's work environment; A target can be defined to generate a healthy campus and to make future engineers aware about people's health in their working environments;
- the need for new organisational forms of technical universities (e.g. for communication with outside of universities through technology transfer agencies, (Filho 2005) or "Forumsakademie" (Narodoslawsky 2006)).

Diverse needs might sometimes lead to measures that are not necessarily compatible. The mismatch might happen through incompatibility of measures, lack of resources for implementing all measures, etc. The task of target setting for systems such as universities with diverse objectives and needs, the development of measures and the assessment of results are therefore not only sequential but they also have to be recursive (see Figure 6).

Figure 6: Relevant activity fields for the recursive decision-making process on objectives and measures

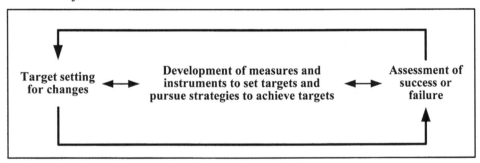

This recursive process is integrated in the social development of the university and its surroundings. One of the most important needs is identified as internal and external communication. Measures for a better communication with the surroundings cannot be developed in an isolated manner at the university alone. The university management needs to consider the relevant inside and outside conditions and include the interests of all involved parties. The socio-technical system should also be able or be prepared for engagement in an intensive and

104 (Shaw et al. 2007).

deep communication process with the university. The interrelation of the technical universities with the social development and its interdependency with new role of technology and new cultural and organizational needs (e.g. understanding of technology in a broader context[105]) will be reviewed in the next chapter.

It should be mentioned again that the examples introduced in this chapter are a sample of existing and possible needs and measures. In the next chapter the author will use these examples to formulate recommendations for the transformation of technical universities for addressing social needs in the 21st century.

---

[105] (Pacey 2004).

10. A participatory approach for technical education in the future

This book presents an analysis on challenges to engineers and technical education in the 21st century. Environmental, economic and social challenges in the future require changes in technical education to accommodate diverse interests of a broad spectrum of involved actors. There have been various ideas and objectives, concepts and measures proposed for modification of technical education (Chapter 8 and 9). Not only future technical education goals and measures are diverse but also the actor groups involved in the decision-making and assessment process vary substantially both culturally and geographically.

This book has shown examples of the influence of social, political, cultural and economical factors on technology development in a socio-technical system and various expectations from the engineering work and technical education. The debates on the most appropriate "strategy for change" seem to happen in a fragmented way, because of:

- isolated discussions on education, research and organisation objectives of universities;
- rare and isolated discussions on technical education, engineering role and technology development objectives;
- no serious discussions on impacts of individual measures on education, research and organization quality;
- rare discussions on interactions among local, regional, national and international strategies and their impacts.

Two preliminary recommendations are presented in this chapter addressing these problems and the need for a broad social debate on the changes of technical education in the context of technology development and engineering work in a socio-technical system. These two recommendations are related to the need for comprehensive debates and improvement of understanding of the potential impacts of isolated measures on technical education.

The *first recommendation* calls for integration of public opinion, engineers' ideas, and stakeholders' (non-engineers') interests into a broad social debate. Such social debates should support tradeoffs among divergent objectives and be used as the steering frame of measures. Conducting such debates needs a bridge between the society and the engineering profession; to improve the technology literacy for understanding the relation between technology development, engineering work and technical education.

The *second recommendation* in this book emphasises the need for a better understanding of relations between technology development and social development by engineers and the implication of this reflexive knowledge in both engineering work and technical education.

10.1 First recommendation

> There is a need for communication among the engineering community and technical universities with society to initiate the integrative and substantial changes in technical education.

Social debates are a medium for communication, providing the information essential for decision-making and can be an arena for learning others perspectives. Relevant actors who take part in social debates prepare and provide information and knowledge (to act) and they need to understand and learn new facts and opinions (to learn)[106]. The communication objective is in this case a better understanding of relations among technology development, the engineering profession and technical education. The technology literacy is regarded here as a precondition for participation of the public in social debates. The public needs to be aware of opportunities and risks of technological developments in order to be prepared to discuss them.

Technology literacy and awareness:

Without technology literacy and awareness of the public it is not possible to have a serious social debate on the new role of technology in society. Currently there are some movements at global, local and national levels on public discussions regarding environmental issues and other impacts of technologies. Environmental policies of the United Nations and the European Union have initiated some activities and included a few participative elements in order to inform society about the risks and rewards of technical solutions. The Convention on Biological Diversity[107], the Alpine Convention[108], or the Aahrous Convention[109] on public partici-

---

[106] See (Rip 2006).

[107] "Signed by 150 government leaders at the 1992 Rio Earth Summit, the Convention on Biological Diversity is dedicated to promoting sustainable development. Conceived as a practical tool for translating the principles of Agenda 21 into reality, the Convention recognises that biological diversity is about more than plants, animals and micro organisms and their ecosystems – it is about people and our need for food security, medicines, fresh air and water, shelter, and a clean and healthy environment in which to live." http://www.cbd.int/convention/ default.shtml, accessed 13 May 2008.

[108] "The Alpine Convention is a framework agreement for the protection and sustainable development of the Alpine region. It was signed on November the 7th 1991 in Salzburg (Austria) by Austria, France, Germany, Italy, Switzerland, Liechtenstein and the EU. Slovenia signed the convention on March the 29th 1993 and Monaco became a party on the basis of a separate additional protocol. The Convention entered into force on March the 6th, 1995." http://www.cipra.org/en/alpenkonvention, accessed 13 May 2008.

[109] "The UNECE Convention on Access to Information, Public Participation in Decision-Making and Access to Justice in Environmental Matters was adopted on 25th June 1998 in the Danish city of Aarhus at the Fourth Ministerial Conference in the 'Environment for Europe' process." http://www.unece.org/env/pp/, accessed 13 May 2008. See also (Kroiss 2001).

pation includes participative elements for long-term development planning including technical solutions. Many such initiatives regard environmental protection as a part of basic human rights and ask for peaceful development goals in a democracy not only for present but also for future generations. These conventions form new elements for a social development with an intensive interaction at local and global levels. These conventions call for voluntary measures to set standards for the local natural conditions in order to achieve the global goal in the future.

Do events at local level achieve the objective?

At local level there are already numerous events at technical universities, technical museums or public research institutes to inform society on special aspects of technologies. Lectures and discussions, engineering days at universities, summer schools, exhibitions, etc. have been organised for public understanding of technology or technological literacy. It should be analysed, whether such events contribute to a well-functioned communication among society, technical universities and the relevant engineering institutions providing knowledge on risks and benefits of technology development and application.

The first recommendation describes the need for quantitative and qualitative improvement of such events for integrating the broad spectrum of society with appropriate methods under the debate. The participatory approaches are a key element in this case.

10.2 Second recommendation

The second recommendation calls for improvement of engineers' understanding of the interactions among technology development, science generation and social development. Reflexive knowledge such as technology assessment supports this understanding and improves the quality of the social debates on bridging the gap between technology development, engineering profession and technical education. A brief review of Technology Assessment in this chapter will show the role of Technology Assessment (TA) as an element for reflexivity.

Rip defines TA as a tool for analysing interactions between society and technology (TA is described in Chapter 7). TA has been developed to support policy decision-making and adaptation of policies regarding technical solutions. It addresses underlying aspects of the new role of technology assessment in society, and it analyses profits and risks associated with new technologies beyond the economic dimension. Even though there are different types of TA with different expectations from TA-studies, there exists a common base for technology assessment. Technology assessment reflects the relationship between social values and technical changes. Public and private sectors apply TA-methods at different intensities to plan, evaluate or support their decisions on technology affairs in order to reduce the negative impacts of technical innovations.

Negative impacts of technologies are however not evaluated identically in different societies. Certain phenomena, such as acid rain or stratospheric ozone layer depletion, are broadly accepted as negative impacts of new technologies, while controversial debates remain on issues such as digital divide or release of Gene Manipulated Organisms (GMOs) into the environment.

TA as a reflexive activity in a socio-economic system

TA contributes to research on interactions between society and technology development, to the awareness of society of opportunities and risks of technologies and the development of participative elements that aid the decision making process. It has however not yet covered all relevant interactions that influence technology development. For example, TA considers the engineering profession and technical education only as a source of information. In a TA for supporting the co-evolutionary governance, however, the black box of the engineering community and technical universities should be opened. One approach that could give further insights into the engineering profession is the participatory approach with active participation of engineers, technical universities staff and researchers of technical research institutes together with society.

An important dilemma, which should be mentioned here, is the complexity of the socio-technical systems which influences the communication between engineers, technical universities and society.

To manage this complexity and sustain the adaptive capacity during periods of change and transitions Rammel (2004) suggests that
- diversity of options and
- combination of different knowledge systems

provide institutional flexibility to cope with surprises and uncertainties.

The complex communication among relevant actors should be well-planned with respect to structure, dynamics, and content of debates. These relevant actor groups for the context of technical education are not only large companies but also public policy figures, entrepreneurs and science and media representatives at local, national and international levels.

The definition of a responsible community for the participatory decision-making with social debates on changing activities discussed in this chapter is based on socio-political dimension of sustainable development and the role of science in sustainable development. Albert (2001) discussed the role of science as a participant in the discussions on the future options regarding the development. Scientists provide knowledge for discussion and at the same time they learn about the needs and insights of the society. The socio-political dimension of sustainable development implies a dual role of actors in the communication process for decision-making, namely to contribute to the learning process of other actors as

172

well as to learn from others. The responsible community for the participatory decision-making should use certain rules for the integration of debates results into a rational decision-making process.

10.3 Involved groups of actors

Based on Albert (2001) the responsible community for debates on science and society in sustainable development should include the following groups:
- Scientific community (including engineering sciences);
- Policy makers;
- Administration;
- Non-Governmental Organizations (NGOs);
- Interested public on relevant issues such as education, health, life quality, consume, energy, transport, communication, security, safety, chances of next generations for a productive and healthy life in the natural environment[110], etc.
- Media.

Actors within diverse groups at local and national levels should also communicate to other actors at a global level.

The engineering profession and student communities are two groups which are not explicitly included in the above list and should be integrated through transdisciplinary cooperation.

Dual roles of each group in the process guarantee a balanced and effective communication in the decision-making process. A detailed discussion of the role of the groups is outside the scope of this work. The author will therefore point out only some aspects of this issue.

The Scientific, engineering and education community could be categorised into different groups based on their functional roles in the *science and technology development process* and their roles in the *conservation of available knowledge* and technological systems.

The learning role of the scientific, engineering and educational community is however less frequent. This is one of the main challenges in the 21st century. The scientific, engineering and educational communities need to learn to have a broader understanding in addition to their disciplinary-specific understanding.

---

110 The Rio Declaration on Environment and Development, The United Nations Conference on Environment and Development, Having met at Rio de Janeiro from 3 to 14 June 1992, Principle 1 out of 27 principles: Human beings are at the centre of concerns for sustainable development. They are entitled to a healthy and productive life in harmony with nature. Principle 3: The right to development must be fulfilled so as to equitably meet developmental and environmental needs of present and future generations (UNEP 1992).

Scientists, engineers and lecturers who are willing to learn about the relations between the real world problems (and needs) and their scientific and engineering activities could support the communication process in a reflexive process with debates on reforms.

Student community: Students are not only consumers of the university and campuses services as a learning group; they are also providers of information for the universities themselves. They are the future generation of decision-makers, actors who can support scientific discussions and research activities and one important group who connects the university and the society together. The provider role of students is important especially in the case of tutors whose roles needs to be defined more clearly at technical universities.

Policy and administration: One of the challenges for policy and administration in social debates is to achieve clear objectives, while technology, innovation, environment, finance, education, social policy and administration at different policy levels cause dissent and have controversial definitions.

The policy and administration also hold the provider role for integration of results as an input to the democratic decision-making process. Nevertheless, according to the socio-political dimension of sustainable development, policy and administration should not control the social debates and the participatory decision-making process regarding the reforms of technical education. Society should use self-organisation capabilities to arrange the debates with the involvement of policy and administration. This precondition leads to the question of whether actors from policy and administration, who are involved in this communication process, could be the same actors responsible for providing financial resources for the communication process.

Policy and administration should also have a learning role. They need to develop a broader understanding of policy fields and have a deep understanding of interactions among different activity fields which influence the technology development and technical education.

10.3.1 NGOs and interested public

Some Non-Governmental Organisations (NGOs) are global campaigning organisations that act outside the local, national and international political systems for public awareness on special issues such as poverty, family, war, environment degradation, agriculture, climate change, conflict resolution, culture, development, children, education, etc.[111]

---

111 Http://www.un.org/dpi/ngosection/dpingo-directory.asp, accessed 13 May 2008.

Some examples of NGO activity fields in different geographical area are given below:

One of the action fields of Greenpeace International is "Creating a toxin-free future with safer alternatives to hazardous chemicals in today's products and manufacturing."[112]

The Institute for International Cooperation of the German Adult Education Association is an example of an NGO for education which focuses on global and life-long learning.[113]

The European Training and Research Centre for Human Rights and Democracy in Graz/Austria (ETC Graz) has its focus on awareness building on human rights and training.

Many non-governmental organizations are partners of the Department of Public Information (DPI) of the United Nations. Currently there are 1533 NGOs with strong information programs associated with DPI.[114] Some of these organsations receive a part of their financial resources from governments.

An NGO is built for a specific responsibility and needs to focus its activities on awareness building. The provider role of NGOs in the reflexive procedure includes awareness building and preparing special educational material on relevant issues such as technology development and technical education in the 21st century. NGOs that are involved in issues related to the environment, education, development, poverty, energy, climate change, young people and women as well as issues on human rights should be integrated in such forums and discussions. Their roles are interpreted as the dashed lincs in the circle in the Figure 7 and it is business as usual for NGOs.

The learning role of NGOs can be understood as a chance for these organsations to gather new information from other participating groups, such as scientific, engineering and education communities or the interested public.

According to the socio-political dimension of sustainable development the interested public is a key group in the social debates and participatory decision-making process. The contributing role of this group is defined by presentation of their life experiences with issues such as experiences with products, industrial activities, technology impacts, education, etc. They can be neighbours of universities and companies, consumers with special needs or critical opinions on special products, families and friends of students and engineers, people with positive or negative opinions on technical solutions, school teachers and children, people who use public transport, drive their own cars, use bicycles or walk to work, etc.

---

112 Http://www.greenpeace.org/international/about, accessed 13 May 2008.
113 Http://www.iiz-dvv.de/englisch/default.htm, accessed 13 May 2008.
114 Http://www.un.org/dpi/ngosection/about-ngo-assoc.asp, accessed 13 May 2008.

All of these people have different ideas and expectations from products, technical systems, engineering work and technical education and can provide their feedback into the social debates.

The learning role of this group can be interpreted as training in technology literacy. There are examples for informing society and interested people on certain technologies but such activities need to be developed further.

10.3.2 Media

Both the classical media such as books and newspapers and the new media such as the internet, play a key role in the reflexive procedure for changes to technical education in the 21st century. They provide platforms for discussions and information exchange which influence public opinion. It is therefore important for the media to learn their critical role in this essential process for the reform of technical education according to the new role of technology development and the engineering profession. For the media to take an active part in the social debates and participative processes it needs to build a balance between freedom of reporting and understanding the critical role of the reflexive process for future generations. Short-term benefits of more publicity and smattering should be compared to long-term benefits of content and choice of topics. Available journals established for public understanding of science are a part of the learning networks with global responsibilities.

10.4 Integration of knowledge by participation

To be effective different opinions need to be integrated through participative approaches into the decision-making process.

Figure 7 presents participative approaches not only as a principle value (in the upper circle) to improve the quality of life but also as a measure (in the lower circle) to integrate public opinion, and different kinds of knowledge necessary for technical innovations. The presented scheme also shows that technical innovation is dependent on technical and non-technical research, engineering profession as well as reflexive research fields such as TA and EIA as well as public and stakeholders' interests.

Figure 7 is an abstract scheme which includes in addition to participation other key normative elements of sustainable development such as technical, social and organizational innovation, learning networks, global responsibility and local engagement; fairness for future generations. It includes inter-relations among the engineering profession, technical and non-technical research, and technical education to technical innovation. Technology assessment is necessary for generation of reflexive knowledge. Participatory approaches should support the integration of different types of knowledge in the decision-making process for selection of technical innovations.

Figure 7: Schematics of relevant issues for reforms of technical education in a 21st century; a set of relevant issues for co-evolution.

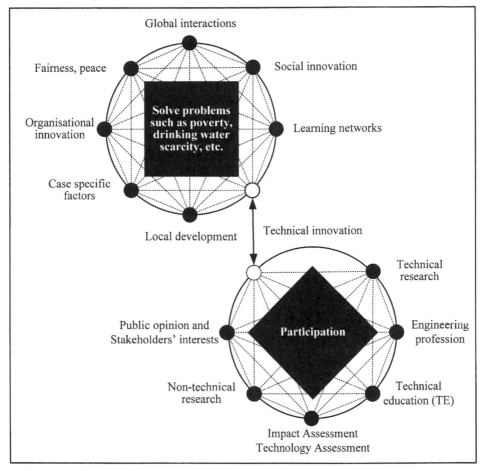

Figure 7 shows that technical universities in sustainable development are involved in a broad scope of activities and trans-disciplinary cooperation to improve the quality of life and protect the environment and health.

Taking into account the technical universities' challenge due to changes in different engineering fields through co-evolution mechanisms, we can assume a high level of internal tension inside universities. Biotechnology, information and material technologies are evolving separately from science and society with different dynamics and regulations. The main challenge to technical education seems to be not the globalisation or environmental problems but the capacity of

177

technical education to deal with the internal conflicts and the continuously changing processes.

The second recommendation in this book emphasises the key role of reflexive knowledge such as TA in engineering work and technical education. This improves awareness about the interactions among technology, science, society and technical universities. The reflexive knowledge such as TA also supports the awareness of involved actors outside of the universities about the new processes and improves their understanding of different possible alternatives in the decision making processes.

10.5 Concluding remarks

The recommendations in this chapter call for social debates on reforms of technical education supported by trans-disciplinarily cooperation with active participation of the public and stakeholders.

The trans-disciplinary cooperation is on one hand necessary in order to know about the average citizens' perception of problems and values and to identify their expectations and to prepare scientific knowledge for public information. (see Kasemir 2003)

On the other hand trans-disciplinary cooperation enables us to promote the dual role of technical universities in providing information and education and to integrate different individuals' knowledge at various positions in the social setting with group-specific languages and different expectations for the reflexive decision-making process.

The framework of the United Nations Decade of Education for Sustainable Development 2004-2015 provides an opportunity to conduct the required trans-disciplinary cooperation in socio-technical system at different geographical and political levels.

Social debates and trans-disciplinary cooperation are preliminary conditions for integrative and substantial reforms of technical education. Debates and cooperation cannot succeed as long as there is a lack of communication due to the technology illiteracy in the society and the absence of engineers' understandings of the interrelation among technical education, technology, science and society. The recommendations presented in this chapter address such deficiencies.

11. Summary of results

The United Nations Decade of Education for Sustainable Development (DESD) 2005-2014 provides an opportunity for awareness building at national and international levels on the requirements for preparing technical education for sustainable development.

The DESD objectives are to improve networking and interactions among stakeholders and to enhance the quality of teaching and learning for sustainable development.

The important areas are providing proper technical education for dealing with complex issues such as climate change, drinking water scarcity or urbanisation that threaten planetary sustainability as well as education to understand principles and values inherent in sustainable development such as human rights, equity, and care for the environment for future generations, biodiversity and global participation.

Some key topics for DESD are therefore as follows:
- new roles of technology development and its new requirements (a precondition is understanding of risks and benefits of technologies);
- new curricula contents of technical education;
- new methods of learning;
- new organisational structures of technical universities enabling new forms of learning and cooperation;
- new national and international regulations for shared responsibilities in presence of uncertainties or high risks;
- new forms of engineering practice;
- more clear definitions for engineers' individual and professional responsibilities allowing them to develop individual profiles that are not totally assimilated in their companies structures.

Last but not least we should be aware of the importance and sensibility of changes in technical universities. Technical universities have a substantial influence on engineers' long careers (about 40 years). The short time in technical education forms the basis for a long-term career. Additionally technical universities are important due to their strong connections to local communities and the real world's complex problems and needs.[115] Technical education could influence

---

115 Each university has its own foundation story. In most cases over time a public and private partnership at local level has generated economic and social stability for the university. The intensity of public or private involvement in decision-making within technical universities induces differences in the quality of engineering education and capabilities of future engineers. This is, however, a short-term and temporary influence. The establishment

engineers' awareness about their responsibilities in trade-off processes. It can contribute towards engineers' learning appropriate knowledge and communication skills for negotiations within organisations[116] in their practice. Technical education has a crucial role in improving the quality of life both in society and for the individual engineers.

Changes in technical universities have long-term impacts that should be considered during the development of new strategies. This book presents results of research on challenges associated with technical education changes in the 21st century towards sustainable development. Two recommendations are made for adapting new strategies (see Chapter 10).

Three key conclusions regarding such challenges can be made from the research and the explorative survey in Part I and II of this book as follows:

1. Massive changes are required in technical education in relation to the future needs of engineering and technology development.

2. Different social values should be taken into account in implementing these changes, since a broad spectrum of actors with very different interests are involved.

3. There is a need for a new and deep understanding of the role of technology in society in order to design harmonic reform strategies for technical universities.

Sections 11.1 to 11.3 will address these challenges. Section 11.4 includes a brief summary of the recommendations suggesting a path for technical education evolution towards sustainable development.

Section 11.5 points out the highlights of future research.

11.1 First key statement: need for massive changes

The analysis has shown a number of reasons for changes to be made in technical education.

The new needs for change in the engineering profession identified in this research are classified into three main categories as follows:

*   *organisational changes* for inter- and trans-disciplinary work on complex problems;
*   *conceptual changes* for the integration of an ex-ante impact assessment of technical design in engineering work;

---

history of technical universities and technical schools shows the strong, long-term effect of local communities in shaping engineering practices. Impulses for changes originate however from the values at global level besides the values of local communities.

[116] (Diani 1986).

- *behavioural changes* to emphasise the individual capabilities of engineers to communicate with the public and an improvement of the self-understanding by individual engineers for their global responsibilities beyond their company and industry rules.

The results show that not only environmental concerns but also business and economic conditions and public perception of the engineering profession imply substantial changes are required in technical education and the engineering profession as summarised in the following sections.

11.1.1 Environmental challenges and engineers' responsibilities

The present role of technical education in the 21st century is shaped by the present situation, past experiences and future perspectives of the role of technology in society. One of the strongest incentives for changing technical education is the negative impacts of technology on the environmentl in the last decades of the 20th century.

In the last decades of the 20th century social debates began to address the negative environmental, social and health impacts of technology and industrial production processes, as well as the products and services themselves. Not only material- and energy-intensive technologies but also service and information technology industries, such as the internet technology and the software industry, have been criticised for their negative environmental, economic and social impacts. An important issue in such discussions has been that people are affected by the environmental hazards or the social and economic pressures associated with technical innovations, regardless of whether or not the impacted individuals are the users of the new products, production processes or services (see Beck 2006).

As a reaction to negative environmental and health impacts of technology, new research fields such as environmental engineering, cleaner production, user interface design and impact assessment have been developed and partly integrated into the engineering education. Concepts such as "zero emission" and "industrial ecology" reflect the desire for an ideal approach towards industrial development and technology application without any negative impacts on humans and the natural environment.

Nevertheless the real world experiences in the past and present have shown that a technical system cannot be implemented and applied without impacts on other systems. This makes the understanding of potential impacts of technical solutions an important element in technical education. Technologies are developed and designed based on certain social values for a specificl user group. Designers can neither practically nor theoretically consider all possible impacts of their works. Risks of malfunctions always remain. Often there are also negative impacts due to the introduction of a new technology interacting with other existing processes or misuse of products (criminal use of human friendly inventions is a prime example of this case). Additionally technologies may evolve after their initial development in various unpredictable directions and might be used in new applications with potentially negative impacts.

Some examples of negative and unwanted side effects of technologies are presented by Porter (1980, P.11) in a list of 45 cases of "social shocks caused by technological developments that went wrong." Some of the famous cases are the thalidomide tragedy[117], the rise and fall of DDT, Asbestos health threat, and human artificial insemination. This list has rapidly expanded and intensified in the last decades. Other examples of technology especially affecting the basic resources of rural people include eutrophication of surface water resources and nitrate contamination of surface and groundwater due to over-fertilization processes. A more global shock is caused by stratospheric ozone layer depletion due to Chlorofluorocarbons (CFC) emissions; accumulation of greenhouse gases in the atmosphere due to the massive use of fossil fuels; Chernobyl tragedy; oil spills and contamination of oceans caused by oil transporters accidents and wars; traffic noise exposure, etc.

Some of the technology-induced problems are not predictable but issues such as oil contamination of oceans or accidents associated with nuclear power plants seem to be considered as acceptable risks today. Such events are only briefly covered in the media, despite the fact that they impact strongly and irreversibly on the environment quality and harm the environment for present and future generations. According to the principles of sustainable development there is a need for more sensitivity and awareness of engineers of such risks in order to care for the quality of life and the environment.

Engineers need to be more conscious about long-range and long-term impacts of their works. Events such as dioxin-contamination of soil near to incineration sites have solely influenced local communities in the long-term, while others such as the Chernobyl catastrophe and the radioactive emissions aftermath contaminated Asian and European continents and possibly beyond. Damages to the natural environment have had long-term effects in both of these cases.

We should also be aware of the potential, serious social impacts due to undesired side effects of new technologies. Some examples include deteriorated social relationships and digital divide caused by information technology, negative impacts of pervasive computing on human behaviour; and threats of security compromise to database networks with personal and private information.

These examples show the vulnerability of daily life to direct and indirect, and short and long-term technological impacts. The sources of hazard or stress can

---

117 Glossary.

182

be physically close or far from the affected people and their lives. Engineers' responsibility is a necessary factor but not sufficient for dealing with the risks of technical innovations.

This leads to the need for the participation of different groups within society including engineers in the decision making process of technology development and applications.

Engineers are a key group to take part in the trade-off evaluation processes and provide information and expertise. Understanding of environmental challenges implies changes in many engineering education fields such as polymer, computer, electrical or mechanical engineering.

11.1.2 Economic and business challenges

The results of the research and explorative surveys in Part I and II of this book indicate that engineers perceive the impact of business on their profession to have increased in the last decades. Working in large international enterprises puts pressure on engineers to have broader knowledge and to be able to expand their job skills within the company. Specialised and narrowly focused qualifications are still important; they are however not considered as a guarantee for a mid- or long-term job opportunity. Another phenomenon, caused by new forms of business, is the wave of establishment of new small companies that work flexibly together on specific projects. Engineers who are involved in such networks need good communication skills, business, law and strategic knowledge as well as a life-long learning program to keep them up to date with the new technologies.

The explorative survey showed that management and business skills are so important that they can be considered as new basic engineering skills. One reason for focusing on hands-on projects in technical education is to prepare engineers for real world business challenges. Inter- and trans-disciplinary working skills are regarded as important new skills for engineers that should be learned during their education.

Respondents of the explorative survey expressed the need for addressing the economic and business challenges in technical education. There were also responses addressing the conflicts between economic and environmental challenges to engineering work and the need for reducing the focus of technical education on the industrial economic imperatives alone.

The third identified challenge to engineering work is the need for changes in technical education to minimise the potential negative public perception of the engineering profession due to negative impacts of technology development and application.

11.1.3 Challenges related to negative image of engineering profession

Public perception of the engineering profession and its work is governed strongly by the satisfaction that engineers provide to society. Negative impacts of technical solutions can lead to users' dissatisfaction. The dissatisfaction can deepen through profound changes in social values by some technology developments and consequently leads to deterioration in the image of the engineering profession, industry and other relevant actors. As soon as existing social values of a new technology begin to evolve the role of the engineering profession undergoes changes as well. Such cases are described as follows:

I.   Undesired impacts of technical innovations in a broad field of applications could lead to a general mistrust of society towards technical innovations and the engineering profession.

II. Undesired impacts of technical innovation in some specific cases could lead to partial mistrust of society towards the engineering profession.

III. The engineering profession might show an active role to avoid undesired impacts of technical innovations. The engineers' shared responsibility would be taken for granted in such cases when engineers take action, before action is taken by others. This does not mean that engineers should take reforms into their own hands without collaborating actively with others in identifying the undesired impacts of technical innovations.

IV. In this case technical universities would perform the reforms to guide the engineering profession and to be able to play an active role in the future development of society.

Cases I and II could lead to strong control of engineering design by external technology management. The engineering profession would increasingly become about pure technical work, delegate responsibility to management and would not be considered any longer a profession with a direct social responsibility. Case III and IV imply active changes in the education programs and organisation of universities focusing on preventing the negative impacts of technical solutions.

11.2 Second key statement: need for diverse changes

The second key statement applies when a broad spectrum of actors from very different interest groups call for engineers to take on new responsibilities and various changes have to be implemented in technical education accordingly.

This study has shown that engineering work is performed in different contexts based on different social values. In the context of elite-role, described in chapter 5, technology is used for conserving political power, fulfilling military goals or satisfying very essential and substantial needs of daily life. Chapter 5 gave also examples of other interests for technology development and other contexts such as control of science over society, fulfilment of market needs or the idea of technology as a social process.

These different contexts of technology development and application imply different needs for the engineering profession and technical education that could result in controversial reforms in technical education. In Chapter 4 different ideas from the university lecturers and engineers regarding the teaching and research experiences during technical education were presented.

The discussions in Parts I and II also showed that social values may change through the emergence of new needs and adverse effects of technical solutions. New social values may lead to new social obligations changing the technology development conditions and requirements on engineering tasks.

The diversity of opinions and positions on technical solutions lead directly to different ideas about the contents, forms and organisation of technical education. These opinions from a broad spectrum of actors should be considered in discussions on changes of technical education. In the next section the author will focus on one such context for technology development that considers technology development as a social process. The next section shows a brief analysis of different factors for changes of engineering education within this context.

11.2.1 Different social values as a central factor for changes in technical education

The desk research showed that social values address the engineering profession while the survey presented the influence of social values on individual engineers. Engineers' responsibility regarding the impacts of technical solutions is a social value with high relevance both for the companies and individual engineers. Accordingly engineers need well defined "responsibility standards" beyond the formal rules and policies of their companies and profession. At the same time they need a shared responsibility with their employers due to a high degree of uncertainty associated with the complex problems and needs. Therefore we can conclude that:

Both the *Engineering profession* and *individual engineers* are influenced by the social values related to the technology development.

Moreover, engineering skills and profession are influenced significantly by technical universities. At the same time engineers and engineering work influence the technical universities. Figure 8 shows direct and indirect interactions between social values and technical education. Social values associated with technology development also determine the importance of technical universities in a society. Through the relation of technical universities to different local, national or international communities each university transpires its own values and missions and influences future engineers' cultures.

185

Figure 8: Technical education has a multifunctional role for practice and research as well as generation of knowledge and skills and is influenced by social values for technology development at different levels

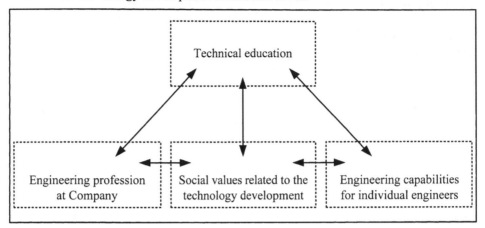

The results of the explorative survey show that technical universities are regarded to influence the engineering profession and technology development through different mechanisms, namely:

- engineers' education,
- generation or reorientation of ideas and world views in the minds of future engineers,
- introduction of engineers to their first practices and their capability to deal with responsibility of engineering work,
- facilitating a discursive arena with society.

The first mechanism is mostly the same at different technical universities. The last three however shape engineering work for various social developments. Different social values, needs of actor groups and impacts associated with the technology development and application constitute various contexts for technology development and engineering work. Each technical university performs in its special context and follows its own implicit and explicit missions and visions and focuses on specific social values.

Engineers who have studied in different universities or lecturers or who have taught at different technical schools might readily recognise the differences among implicit values transferred between different technical universities and their students. The scope and quality of engineering work responsibilities are defined separately by each group and the relation and communication among technical universities and society are formed and regulated differently. Responsibility gains its meaning on the basis of accepted values in a society. Although sustainable development has its own global and general principles such as human rights which should be considered in each case, it calls for the emphasis on the importance of case-specific cultural and local values. The relevant social values for technical universities in sustainable development should therefore be discussed at the micro level in technical universities, at local and regional as well as at international macro level.

The discussion above shows the central role of different social values in the decision-making mechanism for technology development and in the definition of the different roles of technical universities and technical education (Figure 9).

Figure 9: Direct and indirect interactions of social values with technical education

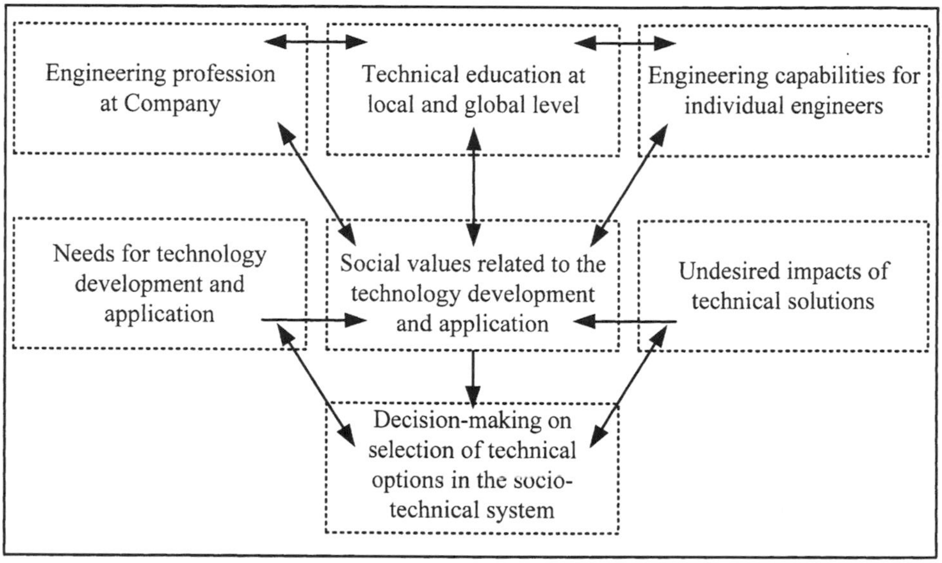

Figure 9 points out the issue of local and global challenges. Technical education, usually at a local level, prepares engineers who may work at a global level in foreign countries with different cultural, social and political environments. The engineers are expected to be able to take different roles for design, management, development and implementation or service of technical facilities abroad. They are also expected to be able to deal with communication, coordination, or participation processes for selection and realisation of solutions that go beyond pure technical activities. The success of incorporating different interests in the discussions on technical education is therefore important for practical reasons.

This section presented arguments for the central role of social values in technical education. In the next section, the author summarises results that address the contents of different interests and social values which influence the engineering profession and its expected role with respect to technology development and application.

187

11.3 Third key statement: new role of technology

> There is a need for a new and in depth understanding of the role of technology in the society and the engineering profession of the 21st century for understanding the new role of technical universities.

Figure 9 in the last section shows the interactions between technical education and social values. The change in the role of technology is shown to be related to the changes in technical education.

The role of technology in the past three centuries can be defined as using, exploiting, manipulating and controlling nature for short-term economic benefits. In the seventeenth century farmers had to have techniques to destroy lay crops and fight animals in order to guarantee the maximum use of natural resources for human beings. (see Bayerl 2001) In 1767 Johann Beckmann offered in his work on natural history a detailed list of natural resources that could be used to fulfil human beings' needs.

This philosophy of using nature as a resource for human beings without reflection of consequences of technical solutions is still alive. There are many existing cases that provide evidence for this philosophy. An example is the production and overuse of aggressive and toxic pesticides and insecticide in the 21st century for maximal use of natural resources despite the negative environmental impacts. Today there are new social developments that require actions to protect nature with or without short-term economic benefits. Examples show that social values for technology development have changed over time due to cultural changes. Today, there are groups of consumers, politicians, engineers and other relevant actors who see products (and technology) in a broader view of a socio-technical system. In Figure 1 the author addressed the complexity that exists even within a simple model of a socio-technical system. This complexity reflects the problem of only focusing on technical and economical optimisation of technical innovations. Moreover there is a need for reflexive knowledge for judging acceptability of designed products and studying long-term impacts of such products on natural habitat.

The socio-technical system is a learning system based on experiments. Selection of needs and evaluation of outputs are based on values, social norms and interactions among different interest groups. Values and interests influence the definition of problems (inputs to a technology development system) and the required solutions (desired situation and results or desired output of a technology development system). They also influence strategies and activities within the system.

Reforms of technical education should consider all these challenges. Lessons from history show that the engineering profession has evolved as societies develop and has also been shaped by individual people during the transition phases.

11.4 Summary of recommendations on a new understanding of engineering profession

In Chapter 10 two recommendations were presented to improve the inclusion of all involved actors in broad social debates and to generate the common understanding of new expectations from engineers and technology development. The first recommendation calls for improvement of technology literacy and the second one recommends the improvement of individual engineers understanding of technology, science and society interdependencies. Understanding of dynamics of these interdependencies is crucial for technical universities in dealing with the development of different engineering fields. The declaration of Barcelona 2004 (chapter 8) indicates the explicit demand on more responsibility for individual engineers by using comprehensive systematic thinking and impact assessment to reduce negative impacts of technologies.

> Both recommendations are based on sustainable development principles participative approaches for inclusion, information exchange and decision-making process.

A community based on this participatory process should be structured based on socio-political dimension of sustainable development and the role of science in sustainable development. Albert et al. (2001) discuss the role of scientists as participants in the discussions on future developments. Scientists provide knowledge in the discussions and at the same time they learn about the needs and insights of society. The socio-political dimension of sustainable development implies a dual role of actors in the communication process for decision-making, namely to contribute to the learning process of other actors as well as learning from others.

The responsible community should integrate the following groups:
* Scientific community including engineering science;
* Policy Makers;
* Administration;
* Non-Governmental Organizations (NGOs) including student associations;
* Engineers and public interested in relevant issues such as education, health, life quality, consumers, energy, public infrastructure, transport, communication, security, safety, next generations' opportunities for a productive and healthy life in the natural environment[118], etc.
* Media (see also Albert et al. 2001).

---

118 The Rio Declaration on Environment and Development, The United Nations Conference on Environment and Development, Having met at Rio de Janeiro from 3 to 14 June 1992, Principle 1 out of 27 principles: Human beings are at the centre of concerns for sustainable development. They are entitled to a healthy and productive life in harmony with nature. Principle 3: The right to development must be fulfilled so as to equitably meet developmental and environmental needs of present and future generations. (UNEP 1992).

Dual roles of each group in the process should guarantee a balanced and effective communication in the decision-making process.

11.5 Further research

This book has emphasised the need for qualitative research for understanding of different terminologies among engineering sciences, natural, social and human science as well as the public's language. This is one of the most important and challenging points for successful inter- and trans-disciplinary cooperation for technical education. In addition different understandings regarding goals of Sustainable Development should be better identified and addressed (see Chapter 4).

Further challenges for engineering that are shortly presented in Chapter 3 and should be addressed in the future are:
- responsibility at local and global level;
- responsibility for health risks in industrial production;
- challenges for female engineers;
- public image of the engineering profession;
- tension between long-term and short-term goals;
- links and tension between engineering and science;
- obstacles to R&D and engineering research.

A number of challenges for technical education were identified with a social process context for technology development in Chapter 6 as follows:
- multiple legitimate needs;
- unintended impacts of technologies;
- trade-off with higher uncertainties;
- challenges due to unexpected changes.

The need for the integration of reflexive knowledge and skills that was introduced in Chapter 7 for decision-making in engineering work should be specified for different engineering fields.

The social obligation of technical universities that was discussed in Chapter 8 brought a number of requirements for universities into the light. These goals should be specified further in details and the list should be expanded by future research.

Chapter 9 shows that some requirements for technical education related to the technical innovation are:
- Better understanding of innovation;
- Learning by doing;
- Understanding of shared responsibility.

Technical universities should therefore apply methods for:

- Integration of universities with their social context;
- Interdisciplinary studies and research;
- Focus on multi-dimensional impact assessment and quality control;
- International cooperation;
- Integration of different interests.

The recommendations for a participative approach for technical education are finally a starting point for building a better understanding between society, the engineering profession and technical universities towards sustainable development. This is the most challenging research issue for the future of the engineering community.

References

Abelson, H., Long, P. D., 2008, MIT's Strategy for educational technology innovation, 1999-2003, *Proceedings of the IEEE* 96(6), 1012-1034.

Akin, W. E., 1977, Technocracy and the American Dream: The Technocrat Movement, 1900-1941, *American history 6(1)*, 104-108.

Albert, R., Brunner, P. H., Fromm, E., Gassner, J., Grabher, A., Kratochvil, R., Krotscheck, C., Lindenthal, T., Milestad, R., Moser, A., Narodoslawsky, M., Paula, M., Rehse, L., Steinmüller, H., Wallner, H. P., Wimmer, R. and Wohlmeyer, H., 2001, *Umsetzung nachhaltiger Entwicklung in Österreich, 2. SUSTAIN Bericht*; Bericht aus Energie- und Umweltforschung, Wien: BMVIT.

Albin, R., 2006, Modern technology as a denaturalizing force, *Poiesis und Praxis (4)*, 289-302.

ANSB, 2005, *Engineering workforce issues and engineering education: what are the linkages?*: American National Science Board, http://www.nsf.gov/nsb/eng_edu/2005_10_20/summary.pdf, accessed on 13 May 2008.

– 2006, *Science and engineering indicators 2006*; Volume 1, Arlington: American National Science Board, National Science Foundation.

Balabanian, N., 2006, On the presumed neutrality of technology, *IEEE Technology and Society Magazine (Winter)*, 15-25.

Bayerl, G., 2001, Die Natur als Warenhaus, Der technisch-ökonomische Blick auf die Natur in der Frühen Neuzeit, in: Hahn, S. and Reith, R. (Eds.): *Umweltgeschichte, Arbeitsfelder, Forschungsansätze, Perspektiven*, Wien: Verlag für Geschichte und Politik Wien, 33-52.

Beck, U., 2006, Reflexive governance: politics in the global risk society, in: Voß, J.-P., Bauknecht, D., Kemp, R. (Ed.): *Reflexive Governance for Sustainable Development*, Edinburgh: Edward Elgar., 31-56.

Beder, S., 1989, Towards a More Representative Engineering Education, *International Journal of Applied Engineering Education 5(2)*, 173-182, http://homepage.mac.com/herinst/sbeder/education2.html, accessed on 13 May 2008.

Bellis, M., 2007, *History of the Railroad*, http://inventors.about.com/od/sstartinventors/a/Stephenson.htm, accessed 13 May 2008.

Blanc, P. D., 2007, *How everyday products make people sick, Toxins at home and in the workplace*, Berkeley and Los Angeles: University of California Press.

Bootsma, M. C. and Driessen, P. P. J., 2005, Exploring a sustainable world, *Committing Universities to Sustainable Development*, Graz, Austria.

Borgmann, A., 2006, Feenberg and the reform of technology, in: Veak, T. J. (Ed.): *Democratizing technology*, New York: State University of New York Press.

Canel, A., 2000, Maintaining the walls: Women engineers at the Ecole Poly-technique Feminie and the Grandes Ecoles in France, in: Canel, A., Oldenziel, R. and Zachmann, K. (Eds.): *Crossing boundaries, building bridges, Comparing the history of women engineers 1870s – 1990s*, Amsterdam: Harwood Academic Publishers, 127-158.

Canel, A., Oldenziel, R. and Zachmann, K. (Eds.), 2000, *Crossing boundaries, building bridges, Comparing the history of women engineers 1870s – 1990s*, Amsterdam: Harwood Academic Publishers.

CEAS-UCB, *Questionnaire: What do you think are the most important skills for the Engineer of 2020?*, 2008], http://engineering.colorado.edu/overview/EAC_Eng2020_Questionnaire.doc, accessed 13 May 2008.

Chandak, S. P., 2006, *Making Right Choices: A Framework for Sustainability Assessment of Technology (SAT): United Nations Environment Programme, Division of Technology, Industry and Economics, International Environmental Technology Centre*, (UNEP DTIE IETC), http://www.etvcanada.ca/forum-2006/SChandak_UNEP2.pdf, accessed 13 May 2008.

Clough, D. W., 2005, *The engineer of 2020 project, A high risk, high-pay-off approach for the future of U.S. engineering education*, http://www.gatech.edu/president/assets/ASEEeng2020.ppt#280,27,Concluding%20remarks, accessed 13May 2008.

COM, 2004, COM(2004) 38 *Stimulating Technologies for Sustainable Development: An Environmental Technologies Action Plan for the European Union*. Commission of the European Communities, http://ec.europa.eu/environment/etap/pdfs/com_2004_etap_en.pdf, Access 13 May 2008.

– 2006, *Women in Science and Technology, the Business Perspective*: EU DG Research, http://ec.europa.eu/research/science-society/pdf/wist_report_ final_ en.pdf, accessed 13 May 2008.

COPERNICUS-CAMPUS, 2007, *University Charter* 2008], web.archive.org/web/20070829025907/http://www.copernicus-campus.org/sites/charter_index1.html.

Cornish Mining World Heritage, 2007, *Matthew Boulton 1728 – 1809,* 2007, http://www.cornish-mining.org.uk/story/boulton.htm, accessed 13 May 2008.

Corso, J. P. D., Ch.Kephaliacos and Marri, M., 2006, Concepts and environmental issues applied to economics and management in the higher technician courses of agricultural education, *EESD 2006*, Lyon.

COUNCIL OF THE EUROPEAN UNION, 2006, *Note: renewed EU Sustainable Development Strategy (EU SDS, access in* 2006], http://register. consilium.europa.eu/pdf/en/06/st10/st10117.en06.pdf, accessed 13 May 2008.

Cross, J., 1998, You say you want a P2 revolution?, *Pollution Prevention Review 8(2)*, 11-25.

Cyrille, H., 2006, Ethics and training: seven pedagogic principles for engineering education for sustainable development, *EESD 2006*, Lyon.

van Dam-Mieras, R. v., 2006, A master sustainable development based on the richness of diversity, *EESD 2006*, Lyon.

Davison, A., 2001, *Technology and the Contested Meanings of Sustainability*, New York: State University of New York Press.

Decker, M. (Ed.), 2001, *Interdisciplinarity in Technology Assessment – Implementation and its Chances and Limits*; in series: Wissenschaftsethik und Technikfolgenbeurteilung, Vol. 11, Berlin/Heidelberg/New York: Springer.

DeSimone, L. D., Popoff, F. and WBCSD (World Business Council for Sustainable Development), 1997, *Eco-efficiency. The business link to sustainable development*, Cambridge, MA et al.: The MIT Press.

Diani, M., 1986, The Social Design of Office Automation, *Design Issues 3(2)*, 73-82, http://links.jstor.org/sici?sici=0747-9360%28198623%293%3A2% 3C73%3ATSDOOA%3E2.0.CO%3B2-J&size=LARGE&origin=JSTOR-enlargePage, accessed 13 May 2008.

Dierkes, M., Hoffmann, U. and Marz, L., 1996, *Visions of technology, Social and institutional factors shaping the development of new technologies*, Frankfurt/Main; New York: Campus Verlag.

Dodds, R. and Venables, R., 2005, *Engineering for Sustainable Development: Guiding Principles*, London: The Royal Academy of Engineering, http://www.raeng.org.uk/education/vps/pdf/Engineering_for_Sustainable_Dev elopment.pdf, accessed 13 May 2008.

Domal, V. and Trevelyan, J., 2006, The Nature of Engineering Work in South Asia: Identifying Engineering Roles, *Technicians Books and conferences Working paper*, http://w3.mech.uwa.edu.au/jpt/eng-work/working-papers/ Trevelyan-Domal-ASEE-2006.pdf, accessed 13 May 2008.

Dougherty, M. R. P. and Hunter, J. E., 2003, Hypothesis generation, probability judgment and individual differences in working memory capacity, *Acta Psychologica 113*, 263-282, http://www.bsos.umd.edu/psyc/dougherty/PDF%20 articles/Doughery%20&%20Hunter,%20Acta%20Psychologica,%202003.pdf, accessed 13 May 2008.

Dyer, M. and Holm, S., 2006, A partnership in environmental technology, *EESD 2006*, Lyon.

Edwards, P. N. and Arbor, A., 1994, From "Impact" to Social Process: Computers in Society and Culture, in: Sheila Jasanoffet al., e. (Ed.): *Handbook of Science and Technology Studies*, Beverly Hills, CA: Sage Publications.

EEA, 2002, *Late lessons from early warnings: The precautionary principle 1896-2000*; Environmental Issue report, Copenhagen: European Environmental Agency, http://reports.eea.eu.int/environmental_issue_report_2001_22/en, accessed 13 May 2008.

El-Sayed, J., 2001, Industrial-Academic Integration takes learning out of the classroom, *New engineering competencies, changing the paradigm!*, Copenhagen.

Emblemsvag, J. and Bras, B., 2000, Process thinking – a new paradigma for science and engineering, *Futures 32*, 635-654.

Encyclopedia Mythica, 2001a, *Prometheus* 2007, http://web.archive.org/web/ 20070612220454/http://www.pantheon.org/articles/p/prometheus.html.

– 2001b, *Tantalus* 2007, http://web.archive.org/web/20070630062313/http:// www.pantheon.org/articles/t/tantalus.html.

European Commission, 2001, Directive 2001/42/EC of the European Parliament and of the Council of 27 June 2001 on the assessment of the effects of certain plans and programmes on the environment.

– 2003, Directive 2003/35/EC of the European Parliament and of the Council of 26 May 2003 providing for public participation in respect of the drawing up of certain plans and programmes relating to the environment and amending with regard to public participation and access to justice, http://eur-lex.europa.eu/LexUriServ/LexUriServ.do?uri=CELEX:32003L0035:EN:HTML, accessed 13 May 2008.

– 2004, COM(2004) 38 *Stimulating Technologies for Sustainable Development: An Environmental Technologies Action Plan for the European Union.* Commission of the European Communities, http://ec.europa.eu/environment/ etap/ pdfs/com_2004_etap_en.pdf, accessed 13 May 2008.

European Environment Agency, 1998, *Europe's Environment: The Second Assessment*, commissioned by: Community, E., Luxembourg: EEA.

Feenberg, A., 1991, *Critical theory of technology*: Oxford University Press, http://www-rohan.sdsu.edu/faculty/feenberg/CRITSAM2.HTM, accessed 13 May 2008.

– 2004, Democratic Rationalization: Technology, Power and Freedom, in: Kaplan, D. M. (Ed.): *Readings in the philosophy of technology*, http://dogma.free.fr/txt/AF_democratic-rationalization.htm, accessed 13 May 2008.

Filho, W. L., 2005, European Reference Point for Technology transfer for Sustainable development: mission and projects, *Committing Universities to Sustainable Development*, Graz, Austria.

Funtowicz, S. and Ravetz, J., 2001, Post-Normal Science – Science and Governance under Conditions of Complexity, in: Decker, M. (Ed.): *Interdisciplinarity in Technology Assessment – Implementation and its Chances and Limits*, Berlin/Heidelberg/New York: Springer, 15-24.

Gassler, H. and Polt, W., 2003, Pfadabhängigkeit, Netzwereffekte und Lock-in. Zur raum-zeitlichen Dimentsion in der ökonomischen Theorie, in: Pichler, R. (Ed.): *Innovationsmuster in der österreichischen Wirtschaftsgeschichte*, Innsbruck, 57-73.

Gilbert, W. and Trudel, P., 2001, *Framing the construction of coaching knowledge in experiential learning theory* 2007, http://web.archive.org/web/20070422155542/http://physed.otago.ac.nz/sosol/v2i1/v2i1s2.htm.

Graaff, E. de (2006). *Psychological aspects of learning and teaching in engineering education*. In 35th international IGIP symposium Engineering Education priority for global development (pp. 27-32).

Graaff, E. d., Thijs, W. and Wieringa, P., 2001, Research as learning paradigm, *New engineering competencies, changing the paradigm!*, Copenhagen.

Grunwald, A. (Ed.), 2002, *Technikgestaltung für eine nachhaltige Entwicklung. Von der Konzeption zur Umsetzung*; in series: Global zukunftsfähige Entwicklung – Perspektiven für Deutschland, Vol. 4, ed. by sigma, Berlin: sigma.

Guston, D. H. and Sarewitz, D., 2002, Real-time technology assessment, *Technology in Society 24*, 93-109.

Hansen, S., 2001, Project assessment as an integrated part of the learning process in the problem based and project oriented study at Aalborg University, *New Engineering Competencies – Changing the Paradigm*, 12-14 September, Copenhagen.

Hill, D. R., 1996, *A History of Engineering in Classical and Medieval Times*: Routledge.

Hill, K., 2005, *Technocracy Movement, An inventory of their finds in The Library of the University of British Columbia Special Collections Division*, http://www.library.ubc.ca/spcoll/AZ/PDF/T/Technocracy.pdf, accessed 13 May 2008.

Hohnen, P. and Potts, J., 2007, *Corporate Social Responsibility, An Implementation Guide for Business*: IISD, http://www.iisd.org/pdf/2007/csr_guide.pdf, accessed 13 May 2008.

Jackson, T. (Ed.), 1993, *Clean production strategies. Developing preventive environmental management in the industrial economy*, 1. edition, Boca Raton et al.: Lewis.

Jischa, M. F., 2004, *Ingenieurwissenschaften*; in series: Studium der Umweltwissenschaften, edited by Brandt, E., Berlin: Springer.

– (Ed.), 2005, *Herausforderung Zukunft Technischer Fortschritt und Globalisierung*, 2. edition, München: Elsevier GmbH.

Joss, S. and Bellucci, S. (Eds.) (Democracy, C. f. t. S. o. and Assessment, S. C. f. T.), 2002, *Participatory Technology Assessment. European Perspectives*, ed. by CSD, London: Centre for the Study of Democracy.

Jucker, H. K., 1998, Chemicals – an Industry in a state of Transition, *Chimia 52*, 147-153.

Kaiser, W. and König, W. (Eds.), 2006, *Geschichte des Ingenieurs, Ein Beruf in sechs Jahrtausenden*, München: Carl Hanser Verlag.

Kasemir, B. et alibi (Ed.), 2003, *Public Participation in Sustainability Science*, Cambridge: Cambridge University Press.

Kemp, R., 2002, Environmental protection through technological regime shifts, in: Bammé, A. e. a. (Ed.): *Technology Studies & Sustainable Development*, München: Profil Verlag GmbH, 95-126.

Kiper, M. and Schütte, V., 1998, Innovation & Zukunftspolitik, *Zukünfte 24 (Summer 98)*, 34-40.

Kohout, F., 1995, *Vorsorge als Prinzip der Umweltpolitik*, München: Tilsner.

Kopfmüller, J., Brandl, V., Jörissen, J., Paetau, M., Banse, G., Coenen, R. and Grunwald, A., 2001, *Nachhaltige Entwicklung integrativ betrachtet, konstitutive Elemente, Regeln, Indikatoren*; in series: Global zukunftsfähige Entwicklung – Perspektiven für Deutschland, Berlin: Edition Sigma.

Koschatzky, K., Kulicke, M. and Zenker, A. (Eds.), 2001, *Innovation Networks*; in series: Technology, Innovation and Policy. ISI Series, Vol. 12, Heidelberg: Physica.

Kroiss, F., 2001, Die Aarhus Konvention. Information und Mitbestimmung in Umweltfragen, *Die Aarhus Konvention. Information und Mitbestimmung in Umweltfragen*, Wien.

Krotschek, C. and Narodoslawsky, M., 1996, The Sustainable Process Index, a new dimension in ecological evaluation, *Ecological Engineering 6*, 241-258.

– 2004, The sustainable process index: a new dimension in ecological evaluation, Ecological Engineering http://www.iisd.org/ic/info/ss9504.htm, accessed 13 May 2008.

LEAS, 2007, *Bleach (sodium hypochlorite)*2007], http://leas.ca/BLEACH.htm.

Lehner, M., 2005, Science-driven vs. market-pioneering high tech: comparative German technology sectors in the late nineteenth and late twentieth centuries, *Industrial and Corporate Chang 14(2)*, 251-278, http://icc.oxfordjournals.org/cgi/content/abstract/14/2/251, accessed 13 May 2008.

Long, G. and Failing, L., 2002, *Sustainability in Professional Engineering and Geoscience: A Primer*: Sustainability Committee of the Association of Professional Engineers and Geoscientists of British Columbia APEGBC, http://www.sustainability.ca/Docs/PrimerPart3MunEng.pdf?CFID=22336896 &CFTOKEN=72887275, accessed 13 May 2008.

Mambrey, P. and Tepper, A., 2000, Technology assessment as Metaphor assessment, Visions guiding the development of information and communications technologies, in: Grin, J. and Grunwald, A. (Eds.): *Vision assessment: shaping technology in 21th century society*: Springer, 33-51.

Marczyk, J., 2000, Uncertainty management and knowledge generation in CAE, *ECCOMAS 2000: European Conference on Computational methods in Applied Sciences and Engineering*, 11-14 September, Barcelona.

McCarthy, T., 1989, *Thomas McCarthy Kritik der Verständigungsverhältnisse zur Theorie von Jürgen Habermas*; in series: Suhrkamp taschenbuch wissenschaft, Vol. 782, Frankfurt am Mein: Suhrkamp.

Moon, F. C., 2005, *Cornell Reuleaux Collection of Kinematic Models*2007], http://web.archive.org/web/20070701014256/http://www.mae.cornell.edu/index .cfm/page/about/reuleaux.htm.

Moyle, P., 2004, *Essays on Wildlife Conservation: Chapter 9: Conservation in the USA: legislative milestones* 2007, http://marinebio.org/Oceans/Conservation/Moyle/ch9.asp.

Mulder, K., 2006, Don't preach. Practice!, *EESD 2006*, 4-6 October, Lyon.

NAE, 2005, *The Engineer of 2020, Project prospectus*; PROJECT SUMMARY for the first Phase of Educating the Engineer of 2020: Adapting Engineering Education to the New Century (2005), Washington: National Academy of

Engineering, Committee on Engineering Education, http://www.nae.edu/nae/engeducom.nsf/0754c87f163f599e85256cca00588f49/85256cfb004a463885256ed800550afd/$FILE/2020%20Prospectus.pdf, accessed 13 May 2008.

Narodoslawsky, M., 2006, Technical universities as platforms for the societal discourse about sustainable development – The case of the "Forumakademie" at the Graz University of Technology, *EESD 2006*, Lyon.

Nentwich, M., Bütschi, D., 2000, The role of participatory technology assessment in policy-making, in: Klein, J. T., Grossenbacher-Mansuy, W., Häberli, R., Bill, A., Scholz, R. W., Welti, M. (Eds.): *Transdisciplinarity: Joint Problem Solving among Science, Technology and Society. An Effective Way for Managing Complexity*, Basel, Boston, Berlin: Birkhäuser.

Nichol, E., 2006, How do engineering educators understand sustainability?, *EESD 2006*, Lyon.

Noggler, L., 2001, Die Wahrnehmung von Luft, das Beispiel einer kleinen Stadt im 19. Jahrhundert, in: Hahn, S. and Reith, R. (Eds.): *Umweltgeschichte, Arbeitsfelder, Forschungsansätze, Perspektiven*, Wien: Verlag für Geschichte und Politik Wien, 121-138.

O'Connell, B. M., 2006, The future of SSIT and the state of engineering profession, *IEEE Technology and Society Magazine (Winter)*, 40-42.

Oldenziel, R., 2000, Multiple-Entry Visa: Gender and Engineering in the US, 1870-1945, in: Canel, A., Oldenziel, R. and Zachmann, K. (Eds.): *Crossing boundaries, building bridges, Comparing the history of women engineers 1870s – 1990s*, Amsterdam: Harwood Academic Publishers, 11-49.

Pacey, A., 2004, The culture of technology, in: Kaplan, D. M. (Ed.): *Readings in the philosophy of technology*: ROWMAN & LITTLEFIELD PUBLISHERS; INC:, 95-102.

Philippi, T., 2005, Lebensphasen-Modelle statt Rollenbilder, von Beamer durch den öffentlichen und privaten Raum, in: Mahrer, H. (Ed.): *Österreich 2050*, 43-56.

Pinter, L., Hardi, P. and Bartelmus, P., 2005, *Sustainable Development Indicators, PROPOSALS FOR A WAY FORWARD*: IISD, http://www.iisd.org/pdf/2005/measure_indicators_sd_way_forward.pdf, accessed 13 May 2008.

Plunket, A., Voisin, C. and Bellon, B., 2002, Research on inter-firm collaboration, evolution and perspectives, in: Plunket, A., Voisin, C. and Bellon, B. (Eds.): *The dynamics of industrial collaboration*: Edward Elgar Publishing Limited.

van de Poel, I. and Verbeek, P.-P., 2006, Editorial: Ethics and engineering design, *Science, Technology, & Human Values 31(3)*, 223-236.

Porter, A. L., Rossini, F. A., Carpenter, S. R. and Roper, A. T., 1980, *A guide-book for technology assessment and impact analysis*; in series: System science and engineering, Vol. 4, edited by Sage, A. P., New York et al.: North-Holland.

Purgathofer, P., 2006, Is informatics a design discipline?, *Poiesis und Praxis (4)*, 303-314, http://igw.tuwien.ac.at/designlehren/purgathofer%20-%20is%20in-formatics%20a%20design%20discipline.pdf.

RAE, 2003, *The future of engineering research*, London: The Royal academy of engineering (Great Britain), http://www.raeng.org.uk/news/publications/ list/ reports/Future_of_Engineering.pdf, accessed 13 May 2008.

Rae, J. and Volti, R., 1999, *The engineer in history*; in series: Worcester Poly-technic Institute, Studies in Science, Technology and culture, Vol. 14, New York: Peter Lang.

Rammel, C., Hinterberger, f. and Becthold, u., 2004, Governing Sustainable Development, A co-evolutionary perspective on transition and change, *GoSD Working papers*, http://www.gosd.net/pdf/gosd-wp1.pdf, accessed 13 May 2008.

Regent University's School of Communication and the Arts, 1997, *Early Philosophers of Technology* 2007, http://web.archive.org/web/2007070408 3557/http://www.regent.edu/acad/schcom/rojc/mdic/early.html.

Rip, A., 2006, A co-evolutionary approach to reflexive governance – and its ironies, in: Voß, J.-P., Bauknecht, D., Kemp, R. (Ed.): *Reflexive Governance for Sustainable Development*, Edinburgh: Edward Elgar.

Rohracher, H., 2000, Acceptance and Appropriation of 'Low-Energy-House'-Components by Users, *International Summer Academy on Technology Studies: Strategies of a Sustainable Product Policy*, July 9-15, Deutschlandsberg, Austria.

Rolt, L. T. C., 1958, *The First Brunel Lecture on I. K. Brunel* 2007, http://web. archive.org/web/20070625135855/http://www.brunel.ac.uk/about/history/ikb/ lecture.

Roman, H. T., 2003, *21 definitions of engineering,* 2007, http://web.archive.org/ web/20060613195201/http://www.uet.edu.pk/ht3/21+Definitions+of+Enginee-ring.htm.

Rosen, P., 2002, Towards sustainable and democratic urban transport: construc-tivism, planning and policy, in: Bammé, A. e. a. (Ed.): *Technology Studies & Sustainable Development*, München: Profil Verlag GmbH, 259-289.

Sanderson, H., Ravetz, J., Tickner, J. and Biddinger, G. R., 2002, *Scientific Application of the Precautionary Principle, (From the September 2002 Globe)* 2007], http:// web.archive.org/ web/ 20060927191521/ http:// www.setac. org/ eraag/era_pp_discourses.htm.

Schäfer, M., Nölting, B. and Illge, L., 2004, Bringing together the concepts of quality of life and sustainability, in: Glatzer, W., von Below, S. and Stoffregen, M. (Eds.): *Challenges for Quality of Life in the Contemporary World*, Dordbrecht/Boston/London: Kluwer, 33-43, http://www.regionalerwohlstand. de/ dwn/236.pdf, accessed 13 May 2008.

Schell, T. v., 2001, Biotechnologie und Gentechnik im Diskurs, in: Skorupinski, B. and Ott, K. (Eds.): *Ethik und Technikfolgenabschätzung, Beiträge zu einem schwierigen Verhältnis*: Helbing&Lichtenhahn.

Schmidt-Bleek, F., 1999, *The Factor 10/MIPS-Concept: Bridging Ecological, Economic, and Social Dimensions with Sustainability Indicators*: United Nations University.

Schot, J. and Geels, F. W., 2007, Niches in evolutionary theories of technical change, A critical survey of the literature, *J Evol Econ 17*, 605–622, www.springerlink.com/content/l1x21235764h2648/fulltext.pdf, accessed 13 May 2008.

SEFI, 2001, New Engineering Competencies – Changing the Paradigm!, *New Engineering Competencies – Changing the Paradigm!*, 12-14-September, Copenhagen.

Shaw, J., Brain, K., Bridger, K., Foreman, J. and Reid, I., 2007, *Embedding widening, Participation and promoting student diversity, What can be learned from a business case approach?*: The Higher Education Academy, http://www.heacademy.ac.uk/assets/York/documents/resources/publications/ embedding_wp_business_case_approach_july07.pdf, accessed 13 May 2008.

Shen, T. T., 1999, *Industrial pollution prevention*; in series: Environmental engineering, edited by Förstner, U., Murphy, R. J. and Rulkens, W. H., 2 edition, Berlin et al.: Springer.

Siegenthaler, C. P. and Margni, M., 2005, *Dissemination, Application and Assessment of LCA in Industry,* 2008, http://www.lcainfo.ch/DF/DF25/ Conference%20Reports.pdf, accessed 13 May 2008.

– 2005a, Dissemination, Application and Assessment of LCA in Industry, *25th LCA Discussion Forum*, ETH Zürich, Switzerland.

– 2005b, *Dissemination, Application and Assessment of LCA in Industry,* 2008, www.lcainfo.ch/DF/DF25/Conference%20Reports.pdf, accessed 13 May 2008.

Siemer, S., Elmer, S. and Rammel, C., 2006, *Pilot study: "Indicators of an education for sustainable development"*; Summary, Vienna: Environmental Protection umbrella Association Austria.

Sotoudeh, M., 2006, Influence of social values on R&D of clean technologies in a sustainable development, *EESD 2006*, Lyon.

Sotoudeh, M. and Mihalyi, B., 2004, Manufactures' response to the needs of users of integrated membrane technology, *Journal of Cleaner Production*, 9.

Sotoudeh, M., Mihalyi, B., Stifter, R. and Siegele, B., 2000, *Bewertung des Durchsetzungspotentials und der Wirtschaftlichkeit vorsorgender Umwelttechnologien, zwei Fallbeispiele*; Endbericht, commissioned by: BMLFUW, November 2000, Wien: ITA.

Spartacus educational, 1999, *Joseph Locke,* 2007, http://web.archive.org/web/19990202154004/http://www.spartacus.schoolnet.co.uk/RAlocke.htm.

Stanford university, 2007, *First high-tech research park, 2007*, http://web.archive.org/ web/ 20070306233523/ http:// www.stanford.edu/ home/ welcome/ research/researchpark.html.

Stöglehner, G., Mitter, H. and Jungmeier, P., 2006, Adult education as a key factor of sustainable rural development, *EESD 2006*, Lyon.

Subai, C., Ferrer-Balas, D., Mulder, K. F. and Moszkowicz, P., 2006, Engineering Education in Sustainable Development, *EESD* 2006, 4-6 Oktober, Lyon.

Sundbo, J., 1998, *The theory of innovation. Entrepreneurs, technology and strategy*; in series: New horizons in the economics of innovation, edited by Christopher, F., Cheltenham, Northampton: Edward Elgar.

Sundermann, K., 1999, Constructive technology assessment, in: Bröchler, S., Simonis, G. and Sundermann, k. (Eds.): *Handbuch Technikfolgenabschätzung*: edition Sigma, 119-128.

Technology Museum of Thessaloniki, 2001, *Ancient Greek Scientists, Mathematician, physicist, engineer Hero of Alexandria,* 2007, http://web.archive.org/web/20070705154740/http://www.tmth.edu.gr/en/aet/5/55.html.

The National Museum of Science and Technology in Stockholm, 2006, *Christopher Polhem,* 2007, http://web.archive.org/web/20061007144608/http://www.tekniskamuseet.se/templates/Page.aspx?id=12313.

The Times, 2001, *ISAMBARD KINGDOM BRUNEL,* 2007, http://web.archive.org/web/20070310230743/http://www.vauxhallsociety.org.uk/Brunel+ Obituary.html.
http://www.vauxhallsociety.org.uk/Brunel%20Obituary.html.

Tichy, G. (Ed.), 2000, *Das Nutzer-Paradoxen und seine Bedeutung für die österreichische Innovationsschwäche*; in series: Wirtschaftsstandort Österreich. Von der Theorie zur Praxis: W. Fuchs und O. Horvath.

Todd, J. A. and Curran, M. A., 1999, *The Streamlined Life-Cycle Assessment*: SETAC North America.

UNEP, 1992, *Rio Declaration on Environment and Development*2007] web.archive.org/web/20070626201643/http://www.unep.org/Documents.multi lingual/Default.asp?DocumentID=78&ArticleID=1163.

University of Florida, 2005, *History of office of sustainability,* 2007, web.archive.org/web/20070325115736/http://www.sustainable.ufl.edu/history. html.

VDI Wissensforum, 2005, VDI Ingenieursstudie Deutschland VDI www.vdi.de/ imperia/md/content/presse/Studie_Wissensforum.pdf.

Veak, T. J. (Ed.), 2006, *Democratizing Technology*, New York: State University of New York.

Vermeulen, W. J. V., 1995, Evaluation in environmental policy as a learning process: the case of PCBs, in: Jänicke, M. and Weidner, H. (Eds.): *Successful environmental policy. A critical evaluation of 24 cases*, Berlin: Sigma, 342-350.

Werk, G. d., Cruz, Y., Rei, J. and Eyto, A. D., 2006, Shaping sustainable professionals, The need to adapt learning processes by involving students, the steps that have to be taken, *EESD 2006*, Lyon.

Wulf, W. A., 1998, Diversity in Engineering, *The Bridge 28(4)*, www.diversity. mtu.edu/diversity_in_engineering.htm, accessed 13 May 2008.

Zabusky, S. E. and Barely, S. R., 1997, "You can't be a stone if you're cement", Re-evaluating the emic identities of scientists in organizations, *Research in Organisational Behaviour 19*, 361-440.

Zimmern, H., 2001, *The translation of The Epic of Shahnameh* Ferdowsi, 2007, http:// web.archive.org/web/20070624092223/http:// www2.enel.ucalgary.ca/ People/far/hobbies/iran/shahnameh.html.

Glossary

The Glossary presents a selected list of key issues to improve understanding of the discussions in the main text.

Abacus

A manual computing device consisting of a frame holding parallel rods strung with movable counters. It is still used to improve the skill of mental calculation and a special form is useful for training blind people.

Black box

An unknown portion of a system: while we know that it exists, we do not know its components. We do, however, know (or at least could speculate on) the inputs and outputs of the black box. Inputs are Resources and entrances available to the system or subsystem; outputs are products or forces that are generated by systems or subsystems.

Bleach

"Bleach, or sodium hypochlorite as it is known chemically, is used in huge quantities around the world, as a disinfectant in municipal waste systems and swimming pools, a laundry whitener and deodorizer in commercial and consumer applications and as a disinfectant used in cleaning. Many schools, care facilities and other institutions specify that bleach must be used routinely as a disinfectant because it is effective on a wider range of bacteria and viruses than many other disinfectants and is cheap and accessible.

Bleach is produced by a chemical reaction that combines sodium hydroxide, water and chlorine – hence the name chlorine bleach. The chlorine in sodium hypochlorite is chemically bonded, but even under the most stringent quality control, the first stage of bleach production, when the chlorine gas is produced, results in the creation of a toxic by product known as dioxin. One of a family of organochlorines, dioxin has been identified as a carcinogen and has been linked to genetic changes and birth defects." (LEAS 2007)

Bologna Declaration 1999[119]

Joint declaration of the European Ministers of Education convened in Bologna on the 19th of June 1999

"… The importance of education and educational cooperation in the development and strengthening of stable, peaceful and democratic societies is univer-

---

[119] Http://ec.europa.eu/education/policies/educ/bologna/bologna.pdf and www.reko.ac.at/universitaetspolitik/dokumente/?ID=681.

sally acknowledged as paramount, the more so in view of the situation in South East Europe.

The Sorbonne declaration of the 25th of May 1998[120], which was underpinned by these considerations, stressed the universities' central role in developing European cultural dimensions.

European higher education institutions, for their part, have accepted the challenge and taken up a main role in constructing the European area of higher education, also in the wake of the fundamental principles laid down in the Bologna Magna Charta Universitatum of 1988. This is of the highest importance, given that universities' independence and autonomy ensure that higher education and research systems continuously adapt to changing needs, society's demands and advances in scientific knowledge.

We consider the following objectives to be of primary relevance in order to establish the European area of higher education and to promote the European system of higher education world-wide:

- Adoption of a system of easily readable and comparable degrees, also through the implementation of the Diploma Supplement, in order to promote European citizens employability and the international competitiveness of the European higher education system;
- Adoption of a system essentially based on two main cycles, undergraduate and graduate. Access to the second cycle shall require successful completion of first cycle studies, lasting a minimum of three years. The degree awarded after the first cycle shall also be relevant to the European labour market as an appropriate level of qualification. The second cycle should lead to the master and/or doctorate degree as in many European countries;
- Establishment of a system of credits – such as in the ECTS system – as a proper means of promoting the most widespread student mobility. Credits could also be acquired in non-higher education contexts, including lifelong learning, provided they are recognised by the receiving universities concerned;
- Promotion of mobility by overcoming obstacles to the effective exercise of free movement with particular attention to;
- for students, access to study and training opportunities and to related services;
- for teachers, researchers and administrative staff, recognition and valorisation of periods spent in a European context researching, teaching and training, without prejudicing their statutory rights;

---

120 Http://www.reko.ac.at/upload/Magna_Charta_Universitatum_Englisch.pdf?PHPSES-
    SID=909dcb513115700adf720221e7f108e6.

- Promotion of European cooperation in quality assurance with a view to developing comparable criteria and methodologies;
- Promotion of the necessary European dimensions in higher education, particularly with regards to curricular development, inter-institutional cooperation, mobility schemes and integrated programmes of study, training and research."

Critical theories of technology

Andrew Feenberg, in Critical Theory of Technology[121], argued that theories of technology fall into one of two major categories: the instrumental theory, and the substantive theory. *The instrumental theory* "offers the most widely accepted view of technology. It is based on the common sense idea that technologies are 'tools' standing ready to serve the purposes of their users. Technology is deemed 'neutral,' without valuative content of its own" (p. 5). Technology is not inherently good or bad, and can be used to whatever political or social ends desired by the person or institution in control. Technology is a "rational entity" and universally applicable, thus allowing similar standards of measure to be applied in diverse situations. Given these propositions, the only response is unreserved commitment to its employment. One may make exceptions on moral grounds, but one must also understand that the "price for the achievement of environmental, ethical, or religious goals ... is reduced efficiency" (p. 6).

In contrast to the instrumental theory *the substantive theory* argues that technology is not simply a means but has become an environment and a way of life: this is its 'substantive' impact". This theory is best known through Ellul and Heidegger. Feenberg continues, "The issue is not that machines have 'taken over,' but that in choosing to use them we make many unwitting cultural choices." (p. 8).

Feedback

Information, signal or measure which tells a person, system or subsystem how well the goal is being achieved, or how well the process is working.

Functional Unit

"The quantity of product that is used to base calculations of material and energy flows across a system." (Todd/Curran 1999)

Impact of Technology

All technologies may have both positive and negative consequences. Engineers must be careful to look for unintended consequences and not just focus on intended ones. Overall judgments about the impact of technology depend

---

121 (Feenberg 1991).

on the weighting of many criteria. Different weightings result in different judgments.

Kyoto Protocol:

KYOTO PROTOCOL TO THE UNITED NATIONS FRAMEWORK CONVENTION ON CLIMATE CHANGE December 1997

It is an amendment to the International Treaty on Climate Change, assigning mandatory emission limitations for the reduction of greenhouse gas emissions to the signatory nations. (Greenhouse gases: carbon dioxide (CO_2), methane (CH_4), nitrous oxide (N_2O), hydrofluorocarbons (HFCs), perfluorocarbons (PFCs), sulphur hexafluoride (SF_6))

Life-Cycle

"Consecutive and interlinked stages of a product system, from raw material acquisition, through manufacturing, use and final disposal." (Todd/Curran 1999)

"The boundaries of the interconnected activities associated with a product or process including all mass and energy inputs and outputs. A system is defined by the function of a product, process, or activity being evaluated." (Todd/Curran 1999)

Life-Cycle Assessment (LCA)

"Compilation and evaluation of the inputs and outputs and the potential environmental impacts of a product or process system throughout its life cycle." (Todd/Curran 1999)

"A phase of LCA involving the accounting of inputs and outputs across a given product or process life-cycle." (Todd/Curran 1999)

Environment is the context that houses a system; also called a supra-system. In this book environment is mostly considered as the physical environment including water, soil and air.

Life-Cycle Impact Assessment (LCIA)

"A phase of LCA aimed at understanding and evaluating the magnitude and significance of the potential environmental impacts of the product or process system." (Todd/Curran 1999)

Life-Cycle Thinking

"Using the life-cycle concept to evaluate environmental issues in a holistic system-wide perspective" (Todd/Curran 1999)

Lowell Statement

The Lowell Statement on Science and the Precautionary Principle is a statement from the International Summit on Science and the Precautionary Principle, hosted by the Lowell Center for Sustainable Production, University of Massachusetts Lowell, September 20-22, 2002. The statement of 85 scientists from 17 countries reads: "We contend that effective implementation of the precautionary principle demands improved scientific methods, and a new interface between science and policy that stresses the continuous updating of knowledge as well as improved communication of risk, certainty, and uncertainty. With these objectives in mind, we call for a re-evaluation of scientific research agendas, funding priorities, science education, and science policy. The ultimate goals of this effort include:

- A more effective linkage between research on hazards and expanded research on primary prevention, safer technological options, and restoration;

- Increased use of interdisciplinary approaches to science and policy, including better integration of qualitative and quantitative data;

- Innovative research methods for analysing the cumulative and interactive effects of various hazards to which ecosystems and people are exposed; for examining impacts on populations and systems; and for analyzing the impacts of hazards on vulnerable sub-populations and disproportionately affected communities

- Systems for continuous monitoring and surveillance to avoid unintended consequences of actions, and to identify early warnings of risks; and

- More comprehensive techniques for analysing and communicating potential hazards and uncertainties (what is known, not known, and can be known)."[122]

Paradigm

"The paradigm is a shared system of interpretation, based on research that is as objective and verifiable as it has been possible to achieve in the area in question" (Sundbo 1998, P. 9).

"According to Thomas S. Kuhn (1967), paradigms are long-term orientation patterns of science, such as classical Newtonian mechanics, that were for long periods unquestioned and were reused and proved again and again by "normal" science, as Kuhn calls it." (Mambrey/Tepper 2000, P. 36)

The concept tries to describe the knowledge generation process and it is criticized by scientists, among them Popper, for its linear and simplistic approach in describing the creation of a paradigm from an existing theory by tests, and for convincing a sufficient number of researchers of the empirical superiority of the theory such that they accept its methodological norms. One of the problems of the paradigm concept is its misuse for sub-areas to establish a paradigm based on special interests of a special theory. However the term has been used in many analytical studies on knowledge generation.

---

[122] Http://www.biotech-info.net/final_statement.html.

Physical indicators

Schmidt-Bleek the inventor of the Material per unit Service Index (MIPS) describes the conditions for physical indicators as follows:

"When attempting to develop indicators for describing the ecological stress potential of goods and services, of individuals, firms, enterprises, regions, countries and the world economy as a whole, such measures should meet the following conditions:

1. They must be simple, yet reflecting essential environmental stress factors. They must be scientifically defensible, albeit not scientifically complete;

2. They should be based on characteristics that are common to all processes, goods and services;

3. The selected characteristics should be straightforwardly measurable or calculable, irrespective of geographic locations;

4. Obtaining results with these measures should be cost-effective and timely;

5. The measures should permit the transparent and reproducible estimation of environmental stress potentials of all conceivable plans, processes, goods and services from cradle to grave;

6. Their use should always yield directionally safe answers;

7. They should form a bridge to economic models;

8. They should be acceptable and usable on all levels: locally, regionally and globally. (Schmidt-Bleek 1999).

Some important quantitative aggregated physical indicators which have been developed in the last decades for use at company, local, national and international level are listed below:

Physical indicators: Ecological rucksacks (MIPS), the Material (incl. Energy) Input per unit Service: MIPS (Material input per service unit) provides an indicator to measure the material consumption for a specific product in the entire life-cycle. The availability of data regarding ecological rucksacks is still insufficient for many countries. [123](Krotschek/Narodoslawsky 1996)

Application: products, Related terms: FIPS = the surface (F for "Fläche", a German word for Surface area) coverage per unit service, TOPS = the eco-toxic exposure equivalent per unit service

Physical indicators: Material flow accounting and analysis (MFA): "The analysis is based on the idea that the extraction of raw materials needed for the production and consumption of a specific product does not simply equal the amount of raw material which make up the product. The final products also

---

[123] Http://www.unu.edu/zef/publications_e/ZEF_EN_1999_03_D.pdf.

contain the raw materials and energy needed for their production. In addition, there are raw materials and energy used for the production of the machines and the preliminary products."[124]

Application: Material flow accounting and analysis (MFA) is an internationally standardised method to measure natural resource use related to production and consumption patterns. An increasing number of national statistical offices report the annual material consumption as part of their environmental statistics. MFAs can be performed on the level of single products."[125]

Physical indicators: Sustainable Pressure Index (1991) SPI: "An index based on only two limiting criteria of sustainability: solar energy and availability of land. Terrestrial assimilative capacity is directly related to land area available for this purpose and the authors argue that this supports the idea of expressing sustainability in area units. Area requirements are calculated around a new definition of process that is leading to the provision of certain services. Major calculable components of this general process include raw material input area, energy supply area input, are requirements of process installation and area for sustainable dissipation of products.

Several case studies exist: pulp and paper industry, electronic industry, agriculture, biotechnology, traffic and transport, energy systems; moreover in regional and municipal planning, insulation of family houses, total pressure of life styles (Austrian data)"[126].

Application: Companies, national and regional strategic planning

Problem-based learning

Problem-based and project-based studies are used for knowledge acquisition directly from experiences. Hansen (2001) presents Schön's Experiential Learning Theory: "Schön's (1983; 1987) theory was developed from case studies of professionals in six domains: (a) architecture, (b) psychotherapy, (c) engineering, (d) scientific research, (e) town planning, and (f) business management. His theory has since been extended into the field of education (e.g. Schön, 1991). For Schön, knowledge construction is a process of critical reflection-in- and on-action that is dependent on the element of surprise. If a practitioner's decision leads to the anticipated result, there is no need to critically reflect on underlying theories. When a decision or action leads to an unexpected outcome, however, this stimulates (in some practitioners) a process of critical inquiry. Schön (1983) describes the process as follows:

---

124 http://www.seri.at/index.php?option=com_content&task=view&id=380&Itemid=142.
125 http://www.unu.edu/zef/publications_e/ZEF_EN_1999_03_D.pdf.
126 (Krotscheck/Narodoslawsky 2004).

When a move fails to do what is intended and produces consequences considered on the whole to be undesirable, the inquirer surfaces the theory implicit in the move, criticises it, restructures it, and tests the new theory by inventing a move consistent with it. The learning sequence, initiated by the negation of a move, terminates when new theory leads to a new move which is affirmed. (p. 155, Schön, D.A. (1983). The reflective practitioner: How professionals think in action. New York: Basic Books.)"[127]

Hansen (2001) regards this method as an operative version of Luhman's general theory of operative constructivism: "The basic idea is that students learn by asking good questions to their own learning process, rather than by answering questions asked by a teacher."

Quality management methods

Technology development implies interactions of companies with universities, research institutes, banks and investors, local and national community, suppliers and customers, international suppliers and customers, etc. On the global market companies screen the suppliers with the expectation of maximising their satisfaction with the purchased product, and produce high quality products in a timely and efficient manner. Standardised quality management methods such as ISO 9000 or ISO 14000 for environmental management are important factors at international level for the selection of suppliers. Additionally, local environmental *regulation* as well as international agreements are emerging in the decision making process. *Regulations* are formal interpretations of a law by administrators; usually regulations add detail to the laws, but must operate within the intent of the law. Some regulations are also established by international and/or national standards-setting groups.

Resources

Any energy, material, information, capital or person available to a system or subsystem.

Screening LCA

"An application of LCA used primarily to determine whether additional study is needed and where that study should focus" (Todd/Curran 1999).

Streamlined-LCA

"Identification of elements of an LCA that can be omitted or where generic data can be used without significantly affecting the accuracy of the results" (Todd/Curran 1999).

---

[127] (Gilbert/Trudel 2001).

Talloires declaration (1990)

"Composed in 1990 at an international conference in Talloires, France, this is the first official statement made by university administrators of a commitment to environmental sustainability in higher education. The Talloires Declaration (TD) is a ten-point action plan for incorporating sustainability and environmental literacy in teaching, research, operations and outreach at colleges and universities. It has been signed by over 300 university presidents and chancellors in over 40 countries."[128]

Points of the declaration:

1. Use every opportunity to raise public, government, industry, foundation, and university awareness by openly addressing the urgent need to move toward an environmentally sustainable future.

2. Create an Institutional Culture of Sustainability

3. Educate for Environmentally Responsible Citizenship

4. Foster Environmental Literacy For All

5. Practice Institutional Ecology

6. Involve All Stakeholders

7. Collaborate for Interdisciplinary Approaches

8. Enhance Capacity of Primary and Secondary Schools

9. Broaden Service and Outreach Nationally and Internationally

10. Maintain the Movement.

Thalidomide tragedy

"A small German pharmaceutical company, Chemie Grünenthal, synthesised thalidomide in West Germany in 1953 while searching for an incxpensive method of manufacturing antibiotics from peptides. By heating phthaloyisoglutamine, the company's chief researcher produced phthalimidoglutarimide, which they soon labelled 'thalidomide.' Chemie Grünenthal patented the molecule and began searching for diseases thalidomide could cure. Without enough research and analysis, Grünenthal management distributed the drug in Switzerland and Germany and considered "a nonlethal sedative would have enormous market potential." Thousands of patients were victim of drug nuisance side effects such as neuropathy, constipation, and fatigue."[129]

Over 10 000 malformed babies were born prior to the ban of the drug because their mothers used the drug. There are also reports about malformed children of young men with malformations due to thalidomide.

---

128 Http://www.ulsf.org/programs_talloires.html.
129 Http://en.wikipedia.org/wiki/Thalidomide.

Appendix A

A summary of an internet discussion: Has engineering been changed?[130]

This Appendix contains a summary of a discussion among engineers in an internet forum. The intent is to show different opinions of engineers regarding key engineering challenges. The reader finds here some of the most pertinent engineering challenges with statements particularly related to the role of information technology in engineering.

Start of the discussion: Has engineering been changed?

"The 45 years since I started my apprenticeship as an electrical engineer have seen continuous change. Most of the products I worked with then are museum pieces, even the tools have changed from a drawing board & slide rule to CAD and CAE tools."

"At the bottom line (at least in pipeline construction) 95% of what we do would be very recognizable to an engineer from the '50's or earlier."

"I don't think the basic process has changed. Engineering is still engineering."

"Our current status is a result of standing on the shoulders of those who came before us."

"At least for heavy civil[131] it has changed fundamentally. At one time the structure was the product. Now the process is more important than the product. At one time the designer would develop a complete and buildable design, generally by himself and a few assistants. Now specifications are thick but generic and no one checks to see if they are applicable to the work. The drawings are numerous but conceptual, relying on shop and "working" drawings from the contractors or their vendors for final details. Designers have less and less experience with actual detailed design or the details of construction or construction costs. The promise of CAD and internet collaboration has not been realized. Scheduling programs that can ignore the details of construction develop unrealistic finish dates for the owners. The result has been the increase in litigation."

"In every case one relies on the intuition, knowledge and integrity of the engineer. If the answer doesn't 'feel' right, he has to have the ability to check it, or else to massively over design, I suppose. That hasn't changed in 200 years."

"If by process you mean:

Problem + engineer + money + time = solution

then no it has not changed, but if you meant the system of doing business then I'd wager it has."

---

130 Http://www.eng-tips.com/viewthread.cfm?qid=143784&page=4. About Eng-Tips: Tecumseh Group, Inc. is an independent forum management company. On the TipMaster (www.tipmaster.com) web site professionals of all types can participate in discussions with others in their specific areas of expertise.

131 Heavy and Civil Engineering construction.

"Yes, it has changed greatly. It now allows competent engineers to accomplish much more in much shorter time and present the results in a professional manner."

"Unfortunately the technology has also allowed incompetent engineers to make many more mistakes, make greater mistakes and to make bigger and worse mistakes in a shorter time while still allowing them to present these mistakes in a professional manner."

"The industry of today seems to allow many more individuals to hide their ineptitude and lack of talent behind the facade of a corporation, like passengers along for the ride. The required skills for engineers are lowered. Many issues are compartmentalized and there is a lack of coordination between different elements."

"I think we are starting to look at the wider picture again. It is less acceptable in the company I work for to say that is Mechanical or Software and not my responsibility. Yes I will be expected to work with engineers of other disciplines but I have to own the problem not walk away from it."

"Skill, knowledge and innovation are still required just the drudge work has been made a lot easier in the process. The skills may be different, the amount of knowledge may be greater but the innovation factor still is important."

"I don't think the process has changed, merely the tools."

"In the "olden" days, the engineer designed and oversaw the construction onsite."

"Great Engineers saw a whole project through from start to finish—these days the "greats" wouldn't want to be bothered with it." "Does that make them less of an engineer?"

"If an architect was the only one to build structures, they would all fall down. If an engineer was the only one to build structures, everyone would tear them all down."[132]

"Modern government design/project people are probably more occupied with proving to their political masters that what is being designed and procured is the "right" thing to be done than actually doing a design/management job."

A summary of statements related to the role of information technology:

"I think the process has changed with the advent of computer simulations."

"Adding in design of experiments software allows you to optimize to a solution much quicker than would otherwise be the case."

"A good example is the use of vascular stents in heart surgery. These are now designed using FEA such that they work better in the human body. How do you think you could do that without much trial and error; doesn't bear thinking about really."

"The technology has only helped the process not fundamentally changed anything."

"We make some assumptions then test these assumptions in models. We just have the ability to have computer models rather than doing all the theoretical modelling by pen and paper and we also have the ability to do calculations more accurately and more ability to have the virtual models refined before moving on to physical models and then the actual product."

---

132 This statement is also related to the inter- and trans-disciplinary character of engineering.

Appendix B

Questionnaire for the Survey 2006-2007

Nr.							
1	Dear participant; The attached questionnaire aims to investigate engineers' ideas on "technical universities for education and research in a sustainable development." Results will be used to show geographical, cultural and case specific needs in different technical universities. They will be used within the book project "links between technical education and sustainability". • Question 1,2,3: background information of interviewees • Question 4a – 9b: please select one option for each question • Question 10-11: please write your suggestion • Question 12: general suggestion on the issue "How do engineers think about technical universities?" (trends, …)						

	Name:						

2	Companies/position/since *(start with most recent)*	1.					
		2.					
		others:					

3	Universities where you have studied /fields and years of graduation: *(start with most recent)*	1.					
		2.					
		others:					

| 4a | Have you had off-campus (industry) learning processes during your education at the university? | yes | no | | | | |
| | | ☐ | ☐ | | | | |

| 4b | How strongly should the external learning process be emphasized at technical universities? | very strongly | strongly | weakly | very weakly | not at all | |
| | | ☐ | ☐ | ☐ | ☐ | ☐ | |

| 5a | Have you had interdisciplinary courses and projects during your education? | yes | no | | | | |
| | | ☐ | ☐ | | | | |

| 5b | How strongly should the inter-disciplinary courses and projects be emphasized at technical universities? | very strongly | strongly | weakly | very weakly | not at all | |
| | | ☐ | ☐ | ☐ | ☐ | ☐ | |

| 6a | Have you participated in project initiatives on local issues during your education? | yes | no | | | | |
| | | ☐ | ☐ | | | | |

| 6b | How strongly should projects on local problems be emphasized at technical universities? | very strongly | strongly | weakly | very weakly | not at all | |
| | | ☐ | ☐ | ☐ | ☐ | ☐ | |

7a	Have you participated in international projects on environmental or social issues during your education?	yes	no			
		☐	☐			
7b	How strongly should participation in international projects on environmental or social issues be emphasized at technical universities?	very strongly	strongly	weakly	very weakly	not at all
		☐	☐	☐	☐	☐
8a	Have you been involved in research activities during your education at technical universities?	yes	no			
		☐	☐			
8b	How strongly should students be involved in research at technical universities?	very strongly	strongly	weakly	very weakly	not at all
		☐	☐	☐	☐	☐
9a	Have you been involved in teaching activities during your education at technical universities?	yes	no			
		☐	☐			
9b	How strongly should students be involved in teaching at technical universities? (Tutors,…)	very strongly	strongly	weakly	very weakly	not at all
		☐	☐	☐	☐	☐

10	Please name three skills you have learned at the university which have been important for your work-life – success.	1.
		2.
		1.
		others:
11	Please name three skills which will be important for the work life in the future and should be learned at technical universities	1.
		2.
		1.
		others:
12	Suggestions regarding to engineering for a sustainable development or for future 2020:	

Thank you for your participation

Table 10: Survey results from questions 10 and 11

Nr.	Country	Date of interview or filling the questionnaire	Gender	Background	Important factors for own success	Important factors for the future
1	Austria	November 2006	W	Chemical engineering, Biochemistry, Biotechnology, 8 years of experience	Analytical thinking Trouble shooting Communication	Analytical thinking Networking Trouble shooting
2	Austria	November 2006	W	Chemical engineering, young engineer	Problem-solving skills Persistence Time-management	Communication, presentation and rhetorical skills Basic entrepreneurial skills Project-management skills Team work
3	Austria	November 2006	W	Chemical engineer, seven years of experience	Team work	Language skills Team work Participation in international projects
4	Finland	October 2006	W	Chemical engineer, EESD 2006 participant	Leadership skills Independency Group working Endurance Understanding other structures	Managerial and leadership skills Group working skills Capacity to understand holistic views Social and language skills
5	Poland	October 2006	W	Chemical engineer, EESD 2006 participant	Literature search skills Research management skills Laboratory skills	Literature search skills Laboratory skills Research management
6	Austria	November 2006	M	Chemical engineering at different universities, 20 years of experience	Modelling skills Flexibility and overview Social skills and cultural knowledge	Inter-disciplinary understanding Data analysis skills Management skills Presentation skills
7	Austria	December 2006	M	Chemical engineer, 10 years of experience	Analytical thinking Problem-solving skills	Foreign languages Management skills

218

Nr.	Country	Date of interview or filling the questionnaire	Gender	Background	Important factors for own success	Important factors for the future
8	Austria	October 2006	M	Chemical engineering, 20 years of research and education experience, EESD 2006 Participant	System analysis Problem-solving skills Communication skills Inter-disciplinary discussion skills	Social skills Basic environmental knowledge Language skills
9	Austria	November 2006	M	Chemical engineer, about 5 years of experience	Endurance Business English skills Presentation skills	Team work Holistic view Career planning
10	Austria	November 2006	M	Chemical engineer, 15 years of experience	Scientific skills	Scientific skills Presentation skills Collaboration skills
11	Austria	October 2006	M	Chemical engineer and manager, 30 years of research and education experiences	Basic engineering knowledge Self confidence Communication skills	Basic engineering knowledge Knowledge on interdependencies of different engineering fields Self confidence
12	Netherlands	October 2006	M	Chemical engineer, background also in economics, 25 years of experience, EESD 2006 participant	Structural and model-based thinking Basic engineering skills Systematic working	Firm basic engineering knowledge Trans-disciplinary problem solving Communication skills Future-oriented attitude System analysis
13	Iran	November 2006	M	Chemical/Environmental Engineer	Teaching skills Research practice Team work	Management skills Communication skills Critical-thinking
14	Sweden	November 2006	M	Civil engineer, 30 years of experience	Executing projects Team work Documentation skills	Presentation skills Language skills Global knowledge Team work

Nr.	Country	Gender	Date of interview or filling the questionnaire	Background	Important factors for own success	Important factors for the future
15	USA	M	July 2006	Civil engineer with background in Electronics, Study in different countries, 20 years of experience	Public speaking and presentation, Problem solving, Working independently	Multi-disciplines, Understanding of environmental and energy issues, Understand of global economy and culture
16	USA	M	July 2006	Computer engineer with industrial engineering background	Teaching skills, Research skills, Planning skills	Team work, Management skills, Communication skills
17	USA	M	July 2006	Computer engineer, young engineer	Computer programming skills, Basic technical understanding, Problem-solving skills	Time-management, Organization skills, Communication skills
18	France	W	October 2006	Engineer: Cleaner Technologies (CT), Waste Science and Techniques, eight years of experience, EESD 2006 participant	Independency, Focus and concentration, Group working	Considering environment and life-cycle, Integration of stakeholders, Considering context and problem-based learning
19	USA	M	July 2006	Electronic engineer, Computer science, 20 years of experience	Computer technology, Electrical engineering, Mathematics, Algorithmic analysis	Biotechnology, Environmental engineering, Internet technologies, Social and networking skills
20	USA	W	July 2006	Electronic engineer	–	–
21	USA	M	July 2006	Electronic engineer and instructor, 30 years of experience	Concentration and focus, Communication and presentation skills, Team work	Basic knowledge skills, Technical leadership, Communication at different levels
22	USA	M	July 2006	Electronic engineer, 15 years of experience	Problem-solving, Team work, Life-long learning	Flexibility, Global vision, Career development
23	USA	M	July 2006	Electronic engineer, 20 years of experience	–	–
24	USA	M	July 2006	Electronic engineer and MBA, about 10 years of experience	Learning skills	–

Nr.	Country	Gender	Date of interview or filling the questionnaire	Background	Important factors for own success	Important factors for the future
25	USA	M	July 2006	Electronic engineer	Constructive discussions and feedback Analytical skills Documentation skills	Interdisciplinary discussions Environmental impact assessment skills Cost impact assessment skills
26	USA	M	July 2006	Electronic engineer	Critical thinking Problem-solving skills	Team work Inter-disciplinary work
27	USA	M	July 2006	Electronic engineer, 20 years of experience	Problem analyzing skills Problem-solving skills Independent research	Problem-solving skills Social skills Understanding of business administration
28	USA	M	July 2006	Electronic engineer, study in different countries, young engineer	General problem-solving skills Persistence, patience, perseverance Basic engineering knowledge Inter-disciplinary skills	General overview of technology Communication skills Resources management and planning
29	Canada	W	July 2006	Electronic engineer, study in different countries, young engineer	Presentation skills Problem-solving skills Endurance	Problem-solving skills Multi-tasking Social skills Presentation skills
30	Austria	W	July 2006	Electronic engineer, 20 years of experience	Logical thinking Problem-solving skills	Practical experiences Strong linkage to industry processes
31	USA	M	July 2006	Industrial engineer, computer engineer, study in different countries, 15 years of experience	Programming skills Applied mathematics Practical experiences	Documentation skills Internet- and web-based languages Internet security Project life-cycle
32	USA	M	July 2006	Industrial engineer, applied statistics, 25 years of experience	Data analysis Presentation skills Problem-solving skills	Leadership skills Presentation, writing and rhetoric skills Focus on real-world problems

Nr.	Country	Date of interview or filling the questionnaire	Gender	Background	Important factors for own success	Important factors for the future
33	USA	July 2006	M	Materials engineer, study in different countries, 12 years of experience	Hands-on experience (electronic device processing) Analytical laboratory skills (materials analysis) Basic statistics knowledge Teaching skills	Understanding of theory and practice Diversification of engineering education Business and marketing knowledge Communication skills
34	Austria	November 2006	W	Mechanical and chemical engineer, 10 years of experience	Computer skills Documentation skills Additional knowledge from other engineering fields Literature research skills	Foreign language skills Knowledge of internet technology Basic knowledge of natural sciences Presentation skills
35	Austria	October 2006	M	Mechanical engineer, about 10 years of experience, EESD 2006 participant	Withstanding problems and going on Technical skills Good time-management	Project- and problem-based learning Interdisciplinary work
36	USA	July 2006	M	Mechanical engineer, study in different countries, 23 years of experience	Learning skills Knowledge of the linkage between engineering and economics Inter-disciplinary education	Social responsibility Ethics in technology and business Information technology and computers Implications and impact of technology
37	USA	November 2006	M	Mechanical engineer, study in different countries, 25 years of experience	Computer skills Research skills Engineering thinking	Computer and engineering software Understanding small-scale phenomena Understanding of environmental issues
38	Netherlands	October 2006	W	Natural scientist, Biochemistry, many years experiences, EESD 2006 participant	Good basic knowledge Social skills Reflection Paper writing skills	Good basic knowledge Social skills Inter- and trans-disciplinary skills Information- and communication technology skills

Nr.	Country	Date of interview or filling the questionnaire	Gender	Background	Important factors for own success	Important factors for the future
39	Spain	October 2006	W	Physicist, teacher in a school of architecture, 25 years of experience, EESD 2006 participant	Forming an opinion considering feedback and different options Interdisciplinary work (scientific and technical)	Interdisciplinary work (scientific, technical, economical, social, etc) Knowledge of several languages Abroad experiences
40	USA	July 2006	M	Ph.D. Physics, study in different countries, 20 years of experience	Language skills Computer knowledge Mathematical skills	Computer knowledge Java programming Mathematical skills
41	Austria	October 2006	W	Planning Engineer, young researcher, EESD 2006 participant	Interdisciplinary approach Presentation and mediation skills Problem analysis and idea generation Project management	Presentation and mediation skills Project management Practical application of knowledge Use of new media (video conference, etc)
42	Sweden	November 2006	M	Process Engineering could be also a kind of Chemical engineering	Broad technical knowledge Project management International experience	Combination of technical and social knowledge Creativity, creating innovations Organizational skills (project management)
43	Netherlands	October 2006	W	Science and policy, EESD 2006 participant	—	—
44	USA	July 2006	M	Surveying and Mapping engineer, 22 years of experience	Surveying Map making Computer programming	Computer skills Hands-on experience Communication skills Management skills

Table 11: Survey results from question 12

Nr.	Country	Date of interview or filling the questionnaire	Gender	Background	Comments to 2020 engineering education
1	Austria	November 2006	W	Chemical engineering, Biochemistry, Biotechnology, 8 years of experience	"More practice and industry-oriented, interdisciplinary projects and courses are necessary. Knowledge about intellectual property of the research results will be very important."
2	Austria	November 2006	W	Chemical engineering, young engineer	"more courses in English; more international guest lecturers; better collaborations/exchange between research groups at the same university, in the same country and across borders (abroad); facilitation and encouragement of interdisciplinary and international projects; teaching of some basic economy and management skills; SOFT SKILLS in addition to scientific knowledge; information/showing of all career option and fields upon graduation; more open/global approaches; a certain contact with industry; ultimately more financial means"
3	Austria	November 2006	W	Chemical engineer, seven years experience	–
4	Finland	October 2006	W	Chemical engineer, EESD 2006 participant	"In 2020 engineering education must be able to guide students to even more holistic comprehension of the global phenomena as today. Today's engineering students are taught details. In 2020 the fractioned knowledge "feeding" necessity will have become outdated, whereas different aspects are always taken into account with appreciation"
5	Poland	October 2006	W	Chemical engineer, EESD 2006 participant	"The importance of cybernetics will increase and there will be a need for more software and more mathematical processes"
6	Austria	November 2006	M	Chemical engineering at different universities, 20 years of experience	"Students should be involved in hands-on project works outside of universities after the basic semesters. Basic semesters are the time between the first second to four years of the education. After learning the basic knowledge of engineering in mathematics, physics, mechanics, chemistry, etc. Their work should be induced and evaluated by the university and they should learn to deal with challenges of the real world business. Projects can be performed in research institutes or companies. The project work could be a part of the thesis or it can be similar to practical projects in medicine education. The advantages of the hand-on project work are for companies which use the knowledge capacity and new ideas and for students who learn about working conditions. In addition projects could be a good chance to find job opportunities. An important advantage of such project works is the support of the creative generation of new ideas at universities. The ideas of project can be generated during current research projects. The financial support of these projects could be provided by scholarships."

Nr.	Country	Gender	Date of interview or filling the questionnaire	Background	Comments to 2020 engineering education
7	Austria	M	December 2006	Chemical engineer, 10 years of experience	"a) More practically oriented study b) More engagement in social projects are necessary"
8	Austria	M	October 2006	Chemical engineering, 25 years of experience, EESD 2006 Participant	"Practica in foreign countries"
9	Austria	M	November 2006	Chemical engineer, about 5 years of experience	"Rather answering what should stay the same: Not all scientific research should be for the sake of industry. Not all funds should come from industry. Not all the research fields should be chosen by industry."
10	Austria	M	November 2006	Chemical engineer, 15 years of experience	"There are different opinions about the functions of a technical university in the universities themselves. Some university managers consider research as the main activity and teaching as the by-product of research at a technical university, while other managers regard teaching as the main activity of technical universities. I think that engineering curricula should transfer basic knowledge but it should emphasize on specific needs of each engineering field. Students would in this way be able to start their practical work at the early phase of their education and use their practical experiences in education and research."
11	Austria				"Future engineering education should put more emphasis on marketing issues (benefit to the customer. Marketing is understood here as business commercialisation with responsibility to analyse available solutions for a problem, comparison of the own developed solution with other solutions, select the best available one, regarding its environmental, economic and social impacts). This will often be in conflict with environmental protection issues. This will have to be handled carefully especially in future. The global environmental problems should be addressed during the education program and students should follow international debates and be able to understand the message of conventions such as Kyoto protocol[130]. Kyoto targets are important because they show the need for energy efficiency. Engineers should consider energy saving in all their design works as a necessary dimension for environmental and economic reasons."

130   Glossary.

Nr.	Country	Date of interview or filling the questionnaire	Gender	Background	Comments to 2020 engineering education
12	Netherlands	October 2006	M	Chemical engineer, background also in economics, 25 years of experience, EESD 2006 participant	"Much more emphasis on inter – and transdisciplinary skills and attitudes, integration in societal needs, more emphasis on communication skills. Understanding of sustainable development in the Bachelor phase. More international exchange in learning and training. Strong problem orientation in the Master-phase. Sense for evaluation of "solutions"."
13	Iran	November 2006	M	Chemical/Environmental Engineer	"1. More multi- and interdisciplinary courses in engineering curricula in 2020. 2. Systematic thinking and a holistic approach will be considered in the engineering curricula in 2020. 3. Teaching and learning techniques will be modified in the engineering curricula in 2020."
14	Sweden	November 2006	M	Civil engineer, 30 years of experience	"My education was in the 1960s and I do not know enough of the present education to be able to answer this question. All my working life I have been dealing with questions on environment and the use of resources and I find this very important."
15	USA	July 2006	M	Civil engineer with background in electronics, study in different countries, 20 years of experience	"Universities should advocate multi-disciplinary learning and teach global issues related to health, environment, culture and economy."
16	USA	July 2006	M	Computer engineer with industrial engineering background	"Any sustainable engineering requires planning for growth and maintenance. Planning for growth requires integrating the needs of the society at present and future, and understanding the current and future technologies at hand. Moreover, understanding the technological advances and predicting their trend issue that sustainability can be maintained."
17	USA	July 2006	M	Computer engineer, young engineer	"Discipline and practice of focused attention on solving problems as well as practice of team work and improved soft skills such as effective communications."
18	France	October 2006	W	Engineer: Cleaner Technologies (CT), Waste Science and Techniques, eight years of experience, EESD 2006 participant	–
19	USA	July 2006	M	Electronic engineer, Computer science, 20 years of experience	"Challenge: High Technology fields such as Computers and Electrical Engineering change very fast. Every 3-5 years industry adopts new technologies and it takes significant efforts to keep up with the technology changes. Lifelong learning is required to remain technically competent. Interdisciplinary skill sets help mitigate effect of industry downturns. For example, dot.com burst of 2001 rendered many software engineers jobless. Software engineers who had interdisciplinary skills in electrical engineering could find jobs easily."

Nr.	Country	Date of interview or filling the questionnaire	Gender	Background	Comments to 2020 engineering education
20	USA	July 2006	W	Electronic engineer	–
21	USA	July 2006	M	Electronic engineer and instructor, 30 years of experience	"There is a need for social environmental awareness, finance awareness, resource usage awareness, multidisciplines knowledge, global problem solving capacity, one or two more languages besides English for engineers"
22	USA	July 2006	M	Electronic engineer, 15 years of experience	"I've worked in three continents and I don't think I learned enough in college about human factors, cultural differences in management, even languages. Students should be prepared to work in different places and in different functions during their career."
23	USA	July 2006	M	Electronic engineer, 20 years of experience	–
24	USA	July 2006	M	Electronic engineer and MBA, about ten years of experience	–
25	USA	July 2006	M	Electronic engineer	–
26	USA	July 2006	M	Electronic engineer	–
27	USA	July 2006	M	Electronic engineer, 20 years of experience	–
28	USA	July 2006	M	Electronic engineer, study at different countries, young engineer	–
29	Canada	July 2006	W	Electronic engineer, study in different countries, young engineer	–
30	Austria	July 2006	W	Electronic engineer, 20 years experience	–
31	USA	July 2006	M	Industrial engineer, computer engineer, study at different countries, 15 years experience	"Keep up with the technology as it grows. Stay on top of new trends and development."

Nr.	Country	Date of interview or filling the questionnaire	Gender	Background	Comments to 2020 engineering education
32	USA	July 2006	M	Industrial engineer, applied statistics, 25 years of experience	"One of the major areas that the technical curriculum has missed out on is "Leadership". They don't try to prepare their students to become a leader. They try to focus so much on the technical aspect that they completely ignore the other aspects. I feel that this has hurt me quite a bit in my professional career and advancement. In today's industry, I think it is more important to have leadership skills than technical, and companies are willing to pay high price for a good leader. Most of the companies' top executives have MBA degrees, which send a message that technical people are not as important. For a technically oriented employee to get to that level, they would have to be a top notch in their field, which is unfair. In my experience, people even with advanced degrees and several years of experience feel trapped and stuck in their current position after so many years of service and don't see any opportunities for advancement. On the other hand, I have seen people with little technical background who could give effective talks and presentations in meetings and company gatherings have excelled in their career very quickly and moved to the top of their professional path."
33	USA	July 2006	M	Materials engineer, study in different countries, 12 years of experience	"Participation in technical conferences, seminars etc. will be more important"
34	Austria	November 2006	W	Mechanical and chemical engineer, 10 years of experience	"I consider the basics of natural science and engineering to be most important. Specialization will take place by training on the job. Therefore university should emphasize the basics; after studying the basics there should be a good variety of specialist courses for the students to choose from, but they do not need to specialize in many areas. In these specialist courses not only the chosen subject should be important, but also to teach students how to get started in a new area, how to find information, how to write a scientific paper. Furthermore presentation skills are important and need to be trained. And the English language is very important, as many scientific papers are written in English. More courses could be taught in English. Students should be obliged to work in industry. Studying at a foreign university for one semester should be encouraged. The university staff who love teaching should do so, and those who feel bothered by teaching obligations should not be forced to do so. That would encourage those who teach to make a good job, and make studying more interesting for the students. At the moment some lecturers might be very good scientists but not good teachers. I think by 2020 many courses will be e-learning courses for the students to study from wherever they wish. Thus those courses taught by a real lecturer should also promote soft/social skills. Social engagement of the students with their university should be encouraged; hence a strong alumni club would keep in touch with graduates. These could help to get the industry more involved into engineering education."

Nr.	Country	Date of interview or filling the questionnaire	Gender	Background	Comments to 2020 engineering education
35	Austria	October 2006	M	Mechanical engineer, about 10 years of experience, EESD 2006 participant	"There is a need for integration of partners (university, economy and politics)" "Engineering education should depend on high interest and motivation"
36	USA	July 2006	M	Mechanical engineer, study in different countries, 23 years of experience	"Avoid a (one-dimensional) techno-centric approach to engineering Promote a ""big-picture"" and a system perspective in engineering Ensure a ""global optimization"" strategy and approach, in addition to the typical ""local optimization"" approach in engineering Tie engineering to economics, and in economical evaluations, consider analysis of the ""overall and total cost"" (e.g., in environmental matters, include potential future losses and opportunity costs such as medical costs and lost time/productivity due to negative health effects, future cleanup & restoration costs, lost opportunities, legal costs, etc., as well as potential future gains such as productivity enhancement, health benefits, etc.) Address and teach the relationship of engineering to ethics, society, politics, policy, environment, etc. Similar to business & law schools, develop and teach ""case studies"" covering past engineering successes and failures Teach concepts of sustainable development "
37	USA	November 2006	M	Mechanical engineer, study in different countries, 25 years of experience	"With the increasing role of computer and software in engineering education and practice, there is a danger of having a generation of engineers with shallow knowledge of the theoretical concepts. The engineering students are becoming less enthusiastic about learning the theories every day. It is very important to explore and use the new ways and techniques for teaching such concepts. With the miniaturization of the systems and devices, in addition to integrating the nano-technology courses into the engineering education, more emphasis on teaching the physics and material structure concepts is also essential. In the same way that engineers learn something about economics, they also need to have at least one course about engineering and environment. It is also good if the engineers have experience in more than one engineering field. For example, it is very useful for the mechanical engineering major students if they chose an electrical engineering minor."
38	Netherlands	October 2006	W	Natural scientist, Biochemistry, many years o experiences, EESD 2006 participant	"Presently, the approach to education is disciplinary-oriented and it works from the abstract to the applied level; rationality is far more important than emotion. I would prefer a starting point of societal complexity and from there going to different disciplinary aspects related to societal problems. That would form a starting point for more disciplinary-oriented study; there would be a better balance between ratio and emotion which is important for learning."

Nr.	Country	Date of interview or filling the questionnaire	Gender	Background	Comments to 2020 engineering education
39	Spain	October 2006	W	Physicist, teacher in a school of architecture, 25 years of experience, EESD 2006 participant	"An engineer should be more capable of coping with the full process where he/she is involved. In an economic world should be desirable where an engineer could refuse any "not honourable work" For example it should not be socially acceptable, if an engineer works for a polluting enterprise." "Engineering students should have an obligation to participate in one or two projects with non-engineering colleagues to understand the importance of non-engineering issues."
40	USA	July 2006	M	Ph.D. in Physics, study in different countries, 20 years of experience	–
41	Austria	October 2006	W	Planning Engineer, young researcher, EESD 2006 participant	"There is a need for more cooperation with industry, politician, local decision-makers, all kinds of stakeholders and the public" (PlanningE-A-W-01) "Engineering education should be more interdisciplinary and interactive" (Europe) "There is a need for project-based learning"
42	Sweden	November 2006	M	Process engineering could be also a kind of Chemical engineering	"Even at "theoretical" universities practical experiences are important for learning success. Therefore I think that project oriented working should take a larger part than it does today. It will probably also be more important to create the right networks (i.e. get to know people from business and for international work) already during studies. Hopefully the "basic" technological knowledge will not be forgotten, as knowledge is a good base for innovations. Creating innovations could be more important and systems on how to do that. In general, how to find information (e.g. via personal networks, via internet etc) could become more important."
43	Netherlands	October 2006	W	Science and policy, EESD 2006 participant	"In 2002 there should be more interdisciplinary and more interactive engineering courses. Less old-fashioned lectures, more discussions and group work. Furthermore, the study should be more oriented towards how to reach sustainable development and towards the influence of society on the subject studied and vice versa"
44	USA	July 2006	M	Surveying and mapping Engineer, 22 years of experience	"Colleges should use latest instruments available in the market in order to prepare students for real world challenges and be ready to start their career upon completion their studies"

Appendix C

Networks, clusters, cooperation and integration of actors in the decision-making

Cluster at regional or national level

M. Porter made the term cluster popular in scientific discussions. The question he addresses is "why nations maintain their competitive advantages in determined sectors of the economy over a long period?"[133]

As factors of success, Porter names a strong network of subcontractor and sales relationships, a network of professional education aligned with the demands of the companies, a research network between companies and universities, an extensive offer of specialized services and the support by economic policy and measures of infrastructure. (Gassler/Polt 2003, P. 61)

A cluster is defined as a group of companies in spatial proximity in a determined sector of economy that is linked by reciprocal relations in information, technology or service networks. (Gassler/Polt 2003, P. 61)

Companies that compete against each other benefit from network-externalities in a cluster of similar companies. Other factor include the common use of public goods such as the infrastructure, a common specialized labour market, advantages from a market for special services and products, the transfer of knowledge and experience via moving employees etc. (Gassler/Polt 2003, P. 61)

Examples of successful high-tech clusters: Silicon Valley and Finland

One of the most popular examples of a successful cluster building is the history of the so-called *"Silicon Valley"*. In the 1950s, the Californian valley was a farming community, but was transformed within a few decades into a micro-electronics high-tech region. The history of Silicon Valley also shows that the military and the aerospace industry have generated demand for semiconductor products for a decade. (Gassler/Polt 2003, P. 68)

This is an important issue. With no doubt more effective research on the technology accelerates the innovation process, but in addition supply is an important issue for new technologies.

The case of Finland with its technology-industry cluster around Nokia shows the importance of public action in the innovation process and the construction of a developed technology industry. As a factor for the Finnish success, Himanen and Castells name the welfare state offering a *"public, free and high-quality education system."* It was again the state, advised by the Science and Technology Policy Council that has in-

---

133 (Porter et al. 1980).

creased public investment in research and development from 1 percent of GDP at the beginning of the 1980 to 3.6 percent of GDP by 2001.

Cooperation at policy level (Harmonisation of policies)

Harmonisation of policies implies identification of needs and optimisation of different policies to use synergy effects. Multiple "win-win" situations are the target of policy harmonisation. Nevertheless each harmonisation has a limit for win-win situations. In the real situation there are different controversial options available for technology, innovation, sector-specific and social policies which may influence technology development directly and indirectly. *The following example of a* "cleaner production promotion" *illustrates the complexity of interactions between different policies and shows several challenges for harmonisation of policies:*

In the last decades the design of promotion strategies for "cleaner production" has been developed continuously. "Cleaner production" is a term which was coined by a working group of the United Nations Environment Programs Industry and Environment Office (UNEP/IEO) in 1989. (Jackson 1993)UNEP adopted the concept of cleaner production, using it to refer to goods, processes and services in the framework of sustainable development. "Cleaner production is the continuous application of an integrated preventive environmental strategy to processes, products, and services to increase overall efficiency, and reduce risks for humans and the environment. Cleaner production can be applied to the processes used in any industry, to products themselves and to various services provided in society."[134]

The most important criteria in this concept are the need to look forward and to bring stakeholders together; to prevent future environmental problems and to consider the negative economic value of waste.

A description of cleaner technology found in a review of cleaner production in ECOTEC states: "Cleaner technology usually reduces polluting emissions to all media instead of shunting them from one to the other."[135]

A large number of different external and internal factors in each industrial sector influence the generation of new ideas, design of new solutions and development and implementation of innovative clean technologies.

Generally speaking environmental policy instruments follow environmental goals. They should save the environment as a collective good, pursue individual producers or customers to share the costs of environmental protection and contribute to the prevention and solution of environmental problems. Some instruments have a prohibiting character to prevent impacts and some of them try to change the economic conditions and social behaviour in order to achieve the environmental goals. Active policy instruments should address different needs and problems of the development of environmental technologies.

---

[134] http://www.unepie.org/pc/cp/understanding_cp/home.htm.
[135] http://europa.eu.int/comm/environment/enveco/industry_employment/annex2.pdf.

There have been different strategies in the search for measures supporting environmental technologies in Europe. After a phase of technology push in the eighties, the market pull and consumer needs have become gradually important since the nineties. Nowadays "push and pull mechanisms operate simultaneously" (Kemp 2002). Jackson (1993, P. 315) defines 24 components of a policy to promote clean production through environmental technologies in OECD areas which include both technology push and market pull effect, as well as organisational changes for environmental technologies. Some of these components are

- Long-term plan for the economy, embodying goals,
- Regulatory control and economic incentives which promote clean production (integrative clean technologies) rather than end-of-pipe technologies,
- Phasing out of subsidies or perverse incentives in some areas (e.g., cheap water) ,
- Generally accepted environmental auditing procedures,
- Where feasible, use of voluntary actions to achieve objectives,
- Regular means to monitor progress and report results to public,
- Training of designers and technicians about advantages of clean production and
- Use of clean production in public operations.

The European Union has promoted the harmonisation of the policy design for promotion of "clean technologies" among others by the Communication of the EU Environmental Technologies Action Plan (ETAP) in January 2004.

One of the main challenges of harmonisation of policies is the participative selection of appropriate measures from the pool of different policies.

Integration of stakeholders' visions

Social consensus on the development of technical innovation is obligatory in modern society which is considered within ideal conditions of this book. Integration of stakeholders' vision into the development process should therefore be performed precisely for specific options, although broad and more general formulations will achieve a consensus more easily. The stability of a consensus should have a dynamic character that enables improvement of options and the speed of achieving consensus and encompass the criteria of specificity and stability. Dierkes (1996) discusses the dilemma of the shaping processes as follows. The consensus should be clear, it should be achieved in a short time and it should be stable. The practical evidence shows that all three conditions cannot be achieved simultaneously. The treatment of the climate change problem gives an example for this dilemma. Specific targets for treatment of the climate change have not been achieved yet. Abstract targets could only be achieved with a partial stability. The vision of economic growth of some stakeholder has displaced the involvement of stakeholders who have alternative visions of a life with a fair growth for future generations. A similar situation has occurred for the stratospheric ozone depletion problem in the 1960s. In the absence of certain scientific evidence for ozone depletion caused by Flororhydrocarbons the vision of a modern industry with modern materials limited the chance of future people to expose themselves to the sun in a healthy way without massive usage of sun protection products. Alternative design solutions existed at that time.

Integration of users in design process and selection of options

The design of integrative environmental technologies needs the cooperation of a large group of actors. An example is described here on the development of membrane technology which is illustrated in (Sotoudeh/Mihalyi 2004):

Membrane technology is a separation technology that is applied in medicine, biotechnology, in the colour and textile industry, waste water treatment, fresh water supply technology, the food industry, etc. Substances, which can be treated by membrane units, are: fruit juices, milk, whey, proteins, enzymes, acids, brines and emulsions.

This technology can be used as an additive technology (end-of-pipe) for example for waste water treatment or it can be integrated into the processes for a selective separation of materials. The latter holds potential environmental and economic benefits such as reduced material or energy consumption or reduced usage of hazardous substances. The environmental advantages of ultra-filtration (a type of membrane technology) are lower energy consumption, reduced CO_2 emissions as well as a smoother and more selective separation process.

The optimal technical operation of membrane units depends strongly on the peripheral units for feed conditioning and on the filtration intervals. The unit must therefore be optimised within the main process. The development of complete systems is possible through cooperation of different firms with pioneer users. In the one of the case studies a small filtration and separation technology firm used the plant manufacturing know-how provided by the established manufacturer as well as by the valve technology, software and engineering firms. In some cases, the co-ordination of such cooperation requires organisational paths, which regulate the fixed relations between the plants manufacturer and other firms.[136] The users' confidence, who requires extensive interaction with the established supplier, promotes the cooperation when it comes to implementation tests of new solutions.

Manufacturers, who focus on the customers during their strategic planning, may use the accommodating relations as a motivator for the necessary users' cooperation in favour of new technical systems. In addition to the user/developer communication there are different groups of stakeholders for the technology policy. Public communication is one of important elements of decision making in the planning. This issue is discussed briefly in chapter 10.

---

[136] However, this strategy will not be successful if there is a user paradox. User paradox describes the double role of users with regard to the innovation behaviour of manufacturers. On the one hand, user needs act as a driving force for small changes; on the other hand, they may be a barrier to substantial change, if users refuse them (Tichy 2000). Since new technical solutions for integrated membrane technology usually require a range of moderate to substantial changes, users may refuse the development of the new solutions.

Regional Innovation Networks (RIN)

RIN are one of the important catalysts in the improvement of regional innovation potential. Such networks include market-based and financially remunerated R&D cooperation as well as non-market-related interactions for exchange of ideas and transfer of knowledge. The European Regional Innovation Survey (ERIS) was carried out between 1995 and 1997. Data were collected from 4,200 questionnaires from manufacturing firms, more than 2,500 from service firms and more than 1,900 from research institutes. With regard to innovation networking, it can be concluded from the ERIS data that

- innovation networking plays an important role in the innovation process in all 11 European regions analysed,
- co-operating firms are economically more successful than firms which do not co-operate,
- spatial proximity between the cooperation partners is less important for customer-supplier relations, but plays a distinct role in horizontal networking between firms and research institutes. (Koschatzky et al. 2001)

The analysis shows that small and medium-sized firms need regional innovation networks to access research institutes and the support of their regional environment.

Technology parks and incubators are a good example of the regional networks near to technical universities.

Technology platforms at EU level

One network-like structure for the innovation of technologies that has been developed in the last years is the European Technology Platforms.

34 Technology Platforms have been established between 2004 and 2007 in the European Union[137] is listed below:

- Advanced Engineering Materials and Technologies – EuMaT
- Advisory Council for Aeronautics Research in Europe – ACARE
- Embedded Computing Systems – ARTEMIS
- European Biofuels Technology Platform – Biofuels
- European Construction Technology Platform – ECTP
- European Nanoelectronics Initiative Advisory Council – ENIAC
- European Rail Research Advisory Council – ERRAC
- European Road Transport Research Advisory Council – ERTRAC
- European Space Technology Platform – ESTP
- European Steel Technology Platform – ESTEP
- European Technology Platform for the Electricity Networks of the Future – Smart-Grids
- European Technology Platform for Wind Energy – TPWind
- European Technology Platform on Smart Systems Integration – EPoSS

---

137 http://cordis.europa.eu/technology-platforms/individual_en.html.

- Food for Life – Food
- Forest based sector Technology Platform – Forestry
- Future Manufacturing Technologies – MANUFUTURE
- Future Textiles and Clothing – FTC
- Global Animal Health – GAH
- Hydrogen and Fuel Cell Platform – HFP
- Industrial Safety ETP – Industrial Safety
- Innovative Medicines for Europe – IME
- Integral Satcom Initiative – ISI
- Mobile and Wireless Communications – eMobility
- Nanotechnologies for Medical Applications – NanoMedicine
- Networked and Electronic Media – NEM
- Networked European Software and Services Initiative – NESSI
- Photonics21 – Photonics
- Photovoltaics – Photovoltaics
- Plants for the Future – Plants
- Robotics – EUROP
- Sustainable Chemistry – SusChem
- Water Supply and Sanitation Technology Platform – WSSTP
- Waterborne ETP – Waterborne
- Zero Emission Fossil Fuel Power Plants – ZEP

Following the definition of the European Research Advisory Board, a technology platform is a "major, pan-European, mission-oriented initiative aimed at strengthening Europe's capacity to organise and deliver innovation". The platform is to "bring together relevant stakeholders to identify the innovation challenge, develop the necessary research program and implement the results."

The Technology Platform has to tackle a major European need, challenge or problem, "rather than simply seek to implement a technology". The Technology Platform requires a road map which develops a medium- to long-term vision of what is needed for Europe, a strategy for achieving this vision and meeting the challenges, as well as a detailed action plan. The action plan should not only consist of research activities but also develop and support strategic innovation activities.

Technology Platforms deal with a financially and organisationally sizable activity and rather with a broad set of technologies than with the generation of one specific technology.

It is necessary that the key players are engaged in the platform to bring the technology successfully to the market. The leadership must rest with the stakeholders rather than with the European Commission.

The structure of the technology platform should be flexible and vary from one to another reflecting political, industrial and market structures.

Since platforms have to substantially change established practices, they need political awareness, political support and thus a high-level of public and political visibility.

Besides this general structure, the European Commission has identified key factors for the success of European Technology Platforms. Contrary to the idea of non-hierarchical structures in a network, the report states that "it is essential that European Technology Platforms have strong leadership with the credibility to bring together and mobilise stakeholders." To this end the involvement of industry is important.

Secondly the Technology Platform should be open in order to avoid becoming a "closed shops" vis-à-vis all relevant stakeholders, including notably small and medium-sized enterprises as well as groups representing wider society interests.

A Platform should have the freedom to determine the most appropriate organizational structure.

Platforms should avoid becoming "talking shops" but should have a clear operational focus from an early stage on, so that research activities begin concretely.

The involvement of national authorities in European Technology Platforms is essential.

Furthermore the Platforms should be proactive in identifying sources of financing. They should not only focus on public funding sources but should also identify potential sources of private funding.

Appendix D

Some *integrative voluntary actions* that are relevant for this analysis are:
- Corporate Social Responsibility (CSR)
- Eco-efficiency concept as an integrative voluntary action
- Environmental Management systems
  - The Eco-Management and Audit Scheme (EMAS)

Corporate Social Responsibility (CSR)

"Social responsibility (is the) responsibility of an organisation for the impacts of its decisions and activities on society and the environment through transparent and ethical behaviour that is consistent with sustainable development and the welfare of society; takes into account the expectations of stakeholders; is in compliance with applicable law and consistent with international norms of behaviour; and is integrated throughout the organisation."

Working definition, ISO 26000 Working Group on Social Responsibility, Sydney, February 2007" (Hohnen/Potts 2007)

Social responsibility Standard ISO 26000 is under development. A short description of the evaluation approach is given by Hohnen (2007):

"How to do an evaluation

Drawing on the CSR objectives and indicators, and the information obtained through the verification and reporting process, firms should consider and respond to the following questions:

What worked well? In what areas did the firm meet or exceed targets?

Why did it work well? Were there factors within or outside the firm that helped it meet its targets?

What did not work well? In what areas did the firm not meet its targets?

Why were these areas problematic? Were there factors within or outside the firm that made the process more difficult or created obstacles?

What did the firm learn from this experience? What should continue and what should be done differently?

Drawing on this knowledge, and information concerning new trends, what are the CSR priorities for the firm in the coming year? and

Are there new CSR objectives?

Finally, it is important that firms celebrate their successes. When goals are met and progress is achieved, all parties concerned need to give each other a pat on the back for a job well done!"

Eco-efficiency concept as an integrative voluntary action

"The term 'eco-efficiency' was coined by the World Business Council for Sustainable Development (WBCSD) in its publication "Changing Course" in 1992. It is

238

based on the concept of creating more goods and services while using fewer resources and creating less waste and pollution."[138]

It is also a part of the concept "sustainable production and consumption". Key dimensions of the eco-efficiency as described by the World Business Council of Sustainable Development in (DeSimone et al. 1997) are

- Reduce the material intensity of goods and services
- Reduce the energy intensity of goods and services
- Reduce toxic dispersion
- Enhance material recyclability
- Maximize sustainable use of renewable resources
- Extend product durability
- Increase the service intensity of goods and services.

Eco-efficiency could be understood as a voluntary tool for structural changes (such as "increasing flexibility in organizational structures") to improve the innovation atmosphere in the European industry. Such needs from the basis for required technological and strategic changes such as "Reduction of Resources Consumption and Emissions through Effective Material and Energy Management".

In this context, solutions with problem-solving potential cannot be defined as an individual technology or activity. They are strategies with a mix of short-term as well as long-term measures for only some parts or the whole of the products' life cycle. These strategies consider not only problems directly related to products and processes but also underlying problems. An example of such solutions would seem as follows:

Wastes & emissions as well as energy and material intensity of manufacturing processes could be decreased by "more focus of process engineering on pollution prevention and maintenance management". Excessive application of chemicals would be reduced by "cooperation of chemical industry with customers to manage the final consumption-related problems". "More cooperation between chemical industry and other industries" would support more environmentally sound services. In this manner some environmental impacts of chemicals such as eutrophication and acidification would be reduced as well. The changes and innovations could be supported and accelerated by long-term political and economical goals. Such examples can be illustrated for customer goods and services, such as fertilizing and painting systems as well as detergents.

Impacts of the eco-efficiency concept among others could be:

- Generation of new ideas for products and processes
- Generation of new discussions and networks for eco-efficient industry.

Environmental Management systems

Application of assessment methods and use of quantitative ecological or sustainability indexes alone is no guarantee for improvement of environmental performance of a company. Environmental management systems are needed to integrate both quantita-

---

138 Http://www.bsdglobal.com/tools/bt_eco_eff.asp.

tive and qualitative factors, support the target setting for improvement of environmental performance and control the development of appropriate improvement strategies. Shen (1999) illustrates different pollution prevention strategies and gives examples on end-of-pipe and integrative measures. He suggests a comprehensive pollution prevention feasibility analyses which integrates technical, environmental, economic and institutional feasibility analysis at company level.

The most important European environmental management system is EMAS (the Eco-Management and Audit Scheme). Different European countries have their own management systems. These are combination of economic management, social management and environmental management strategies.

The Eco-Management and Audit Scheme (EMAS)[139]

EMAS allows voluntary participation in an environmental management scheme for organizations operating in the European Union and the European Economic Area. The scheme has been operative since April 1995. It aims at promoting continuous evaluation and improvements in the environmental performance of participating organisations.

History of EMAS

The EMAS Regulation 1836/93 was first introduced in July 1993 as an environmental policy tool devised by the European Commission, in a step towards the Community's goal of sustainable development. The EMAS scheme was opened for voluntary participation by organizations from April 1995 and its scope restricted participation to sites operating industrial activities.

In 1996, the International Environmental Management System Standard, EN ISO 14001 was published and was recognized as a step towards achieving the goals of EMAS. It was also recognised that all sectors have significant impacts on the environment and that the environment will benefit from good environmental management in other sectors as well.

Article 14 of the regulation allowed Member States to extend the scheme to other economic sectors and several Member States took the opportunity to pilot EMAS successfully in those other sectors. Article 20 of Regulation 1836/93 stated that the EMAS scheme had to be reviewed no longer than 5 years after its entry into force.

In 2001 the new Regulation (EC) No 761/2001 was adopted (O.J. L114 , 24.4.2001, p.1). Its main elements are:
- the extension of the scope of EMAS to all sectors of economic activity including local authorities;

---

139 EMAS certification is a voluntary environmental-oriented action. There are a number of management concepts which integrate economic or social aspects more strongly than EMAS. Source of information: ec.europa.eu/environment/emas/tools/faq_en.htm#what.

- the integration of ISO 14001 as the environmental management system required by EMAS, so that progressing from ISO 14001 to EMAS will be smoother and not entail duplication;
- the adoption of a visible and recognisable EMAS logo to allow registered organisations to publicise their participation in EMAS more effectively;
- the involvement of employees in the implementation of EMAS;
- the strengthening of the role of the environmental statement to improve the transparency of communication of environmental performance between registered organisations and their stakeholders and the public; and
- a more thorough consideration of indirect effects including capital investments, administrative and planning decisions, procurement procedures, choice and composition of services (e.g. catering)

The Regulation consists of 18 Articles and 8 Annexes. Unlike other management systems standards the annexes form part of the Regulation – they are not merely informative, the requirements of the annexes must be met.

The main stages involved in achieving EMAS registration are as follows:
- Investigating an organisation's interactions with the environment in an environmental review;
- Establishing an effective environmental management system on the basis of the review aimed at improving the environmental performance of the organisation;
- Communicating environmental performance data in an environmental statement. This statement is verified by a third party to provide external credibility to this performance-reporting element.

Index

abacus 36, 205

black box 142, 150, 172, 205
bleach 47, 205
Bologna Declaration 146, 205

chemical engineering 40, 42, 54, 67–69, 110
civil engineering 37–39, 42, 65
cluster 231
Corporate Social Responsibility (CSR) 238
creativity 20, 50 f., 105, 163
critical theories of technology 102, 207

eco-efficiency 43, 63 f., 238 f.
Eco-Management and Audit Scheme (EMAS) 238, 240
engineering 17
– associations 18, 33, 58, 61, 101
– community 21, 23, 49, 65, 142, 170, 172, 191
– science 39, 42, 53, 132, 139, 173, 189 f.
environmental management 145, 164, 212, 239, 240

feedback 76, 93, 114 f., 161, 176, 207
female engineer 40–42, 49, 80, 82, 190

global responsibility 45–47, 88, 96, 149 f., 160, 165, 176

impact of technology 17, 81, 207
information technology 18, 40, 44, 54, 62, 80, 106, 111, 181 f., 214 f.

integrative
– approach 136
– clean technology 233
– environmental technology 158, 234
– measure 47, 240
– voluntary action 238
interdisciplinary
– courses, learning 36, 69–71, 74, 216
– studies, research 152, 154–156, 191
– understanding, discussion, skills 80, 162, 169 f., 173 f., 178, 183, 189 f.
– work, cooperation 140, 149, 154 f., 159 f.

Kyoto Protocol 26, 208

life-cycle 52, 56, 133, 208
life-cycle assessment 132 f., 208
life-cycle impact assessment 208
life-cycle thinking 52, 134, 208
life-long learning 76, 93, 175, 183
local engagement 45, 96, 160, 176
Lowell Statement 127 f., 208

mechanical engineering 34, 39, 110, 183

network 89, 95, 116, 123, 149, 235

paradigm 65, 78, 209
participatory
– approach 141, 157, 169, 171 f., 176
– decision-making 172–175
– process 129, 189
– technology assessment, TA 139–141

243

performance targets 52

(public) perception 50, 181, 183 f.

physical indicator 135, 210 f.

platform 124, 176, 235–237

precautionary principle 9, 21–25, 27 f., 45 f., 55, 57, 87, 96, 123, 126–128, 139, 196, 209

problem-based learning 70, 95, 211

quality management 44, 81, 132, 139, 156, 162 f., 212

research and development (R&D) 42, 45, 55 f., 110, 123, 125, 140

resources 19, 27, 38, 46 f., 56, 64, 101, 106, 109–115, 120 f., 134, 136, 151–156, 159, 166, 174 f., 182, 188, 202, 239

responsibility 9–11, 19 f., 44–51, 64–66, 76, 83, 86–94, 96–98, 104, 108, 116, 126, 139, 142, 145, 149, 152, 161–165, 175, 183–186, 189 f., 215, 238

Rio Declaration 126, 146

risk 21, 23, 26, 28, 35, 61–63, 83, 105, 109, 112, 127, 132, 139, 193 f., 209

Screening LCA 212

stakeholder 21, 25, 52–56, 66, 95, 107, 123, 137, 140–142, 146–149, 169, 176–179, 213, 232–234, 236–238, 241

streamlined-LCA 212

student 5, 37, 40–43, 46, 51, 61, 67, 71–74, 84, 89 f., 95, 149–157, 159–164, 166, 173–175, 186, 189, 202, 204, 206, 212, 217

Talloires Declaration 145, 213

teacher 69, 149, 175, 206, 212

technocracy 43, 46

thalidomide tragedy 182, 213

trade-off 123, 125 f., 129, 180, 183, 190

trans-disciplinary 43, 47, 55, 57, 142, 215

– approach 139

– engineering education 43

– skills 87, 95, 183

– team 95, 117, 155

– work, cooperation 57–59, 81, 87, 94, 123, 129, 141, 155 f., 165, 173, 177 f., 180, 183, 190

Umweltbildung, Umweltkommunikation und Nachhaltigkeit
Environmental Education, Communication and Sustainability

Herausgegeben von/Edited by Walter Leal Filho

Band 25 Walter Leal Filho / Mario Salomone (eds.): Innovative Approaches to Education for Sustainable Development. 2006.

Band 26 Walter Leal Filho / Franziska Mannke / Philipp Schmidt-Thomé (eds.): Information, Communication and Education on Climate Change – European Perspectives. 2007.

Band 27 Ulisses Miranda Azeiteiro, Fernando Gonçalves, Ruth Pereira, Mário Jorge Pereira, Walter Leal Filho, Fernando Morgado (eds.): Science and Environmental Education. Towards the Integration of Science Education, Experimental Science Activities and Environmental Education. 2008.

Band 28 Walter Leal Filho / Nils Brandt / Dörte Krahn / Ronald Wennersten (eds.): Conflict Resolution in Coastal Zone Management. 2008.

Band 29 Walter Leal Filho / Franziska Mannke (eds.): Interdisciplinary Aspects of Climate Change. 2009.

Band 30 Mahshid Sotoudeh: Technical Education for Sustainability. An Analysis of Needs in the 21st Century. 2009.

www.peterlang.de

Walter Leal Filho (ed.)

Innovation, Education and Communication for Sustainable Development

Frankfurt am Main, Berlin, Bern, Bruxelles, New York, Oxford, Wien, 2006.
763 pp., num. fig. and graphs
Environmental Education, Communication and Sustainability.
Edited by Walter Leal Filho. Vol. 24
ISBN 978-3-631-55644-3 · hardback € 91.20*

Education for Sustainable Development presents a vision of education that seeks to empower people to assume responsibility for creating a sustainable future. There are many different stakeholders in sustainable development whose participation is essential if long-term developments are to be achieved: governments and intergovernmental bodies, the mass media, the civil society and non-governmental organizations, the private sector and formal education institutions (i.e. schools, universities, research and training institutes). But despite the need for concerted efforts, truly cross-sectoral and interdisciplinary projects and practical activities are seldom seen. In addition, there are few publications where the subject matters of education, innovation and communication – as they apply to sustainable development – are approached in an integrated way. This book combines approaches, methods and analyses which illustrate the contribution of innovation, education and communication to the cause of sustainable development. It is prepared as part of the INTERREG IIIB (North Sea) Project "SmartLIFE", a partnership between three EU regions – Germany, Sweden and the UK. It is also a further means of support to the UN Decade of Education for Sustainable Development, since there is a paucity of academic and at the same time practice-related books which go over and above the theory of education and of sustainable development and move on to address the various problems related to it.

Frankfurt am Main · Berlin · Bern · Bruxelles · New York · Oxford · Wien
Distribution: Verlag Peter Lang AG
Moosstr. 1, CH-2542 Pieterlen
Telefax 0041 (0)32/3761727

*The €-price includes German tax rate
Prices are subject to change without notice
Homepage http://www.peterlang.de

Peter Lang · Internationaler Verlag der Wissenschaften